Peter Jäger

**Calibration compendium
Edition 2020**

Calibration terms, basics and normative references

Bibliographische Information der Deutschen Nationalbibliothek:
Die Deutsche Nationalbibliothek verzeichnet diese Publikation
In der Deutschen Nationalbibliographie; detaillierte bibliographische
Daten sind im Internet über http://dnb.dnb.de abrufbar.

© 2019 Jäger, Peter
p.jaeger.metrologie@web.de
Herstellung und Verlag:
BoD – Books on Demand, Norderstedt

ISBN: 978-3-7504- 3603-9

Table of contents

Preamble 8

Definitions 9

Measurement technology in everyday life 10

Metrology – basic categorization 11

Legal metrology 12

Scientific metrology 14

Industrial metrology 16

International SI system 19

      Defining natural constants 21
      SI-Units 22
      National metrology institutes 25

Accreditation 27

      DKD 29
      DAkkS 31
      Accredited calibration laboratory 32

Normative references 37

      ISO DIN EN ISO 9001:2015 40
      IATF 16949 46
      DIN ISO/IEC 17025:2018 51
      DIN ISO/IEC 10012:2004-03 53

The Commercial Measurement System 56

Eichung (Germany) 58

Why calibrate? 60

Table of Contents

    Profile of a measuring device - added value ..66
    Replacement of a measuring device for cost reasons 68
    Change through use 69

Traceability 70

    Documentation 72
    Measurement uncertainty 72
    Calibration hierarchy 74

Measurement uncertainty 78

    Systematic deviations 79
    Random deviations 79
    Messunsicherheit oder Toleranzangabe 87

Calibration results 92

    Calibration certificate 92
    Intervals in a calibration certificate 97
    Statement of conformity 99
    Decision rule - what does that mean? 101
    Must a device be "compliant" ? 105

Need of test & measuring equipment 107

Labeling of test & measurement equipment 113

Determination / adjustment of calibration intervals 123

    Interval adjustment 125
    Evaluation of calibration results: 126
    Prolongation of calibration interval 129
    Intervals in a calibration certificate 131

Table of Contents

Indication of the time of recalibration 132
Start of a calibration interval 134
Interruption of usage 134
General information about calibrations and
calibration intervals 135

Calibration planning/scheduling/delivery 136

Factory or traceable calibration 136
Measuring chain or individual devices 143
Failure of a measuring system / repairs 145
Calibration: Laboratory, „on-site" or „in-situ"
.... 146
Selecting a calibration laboratory 147
Class of calibration 149
After a calibration / receiving your instrument
.... 150

Machine capability 153

Measurement system analysis MSA 156

Type-1 study 157
Type-2 study, Gauge R&R study 158
Type-3 study, R&R study 159

Bibliography 161

Preamble

This book is intended to provide concentrated information on calibration. It is intended to guide the most important standards and the reference points without the reader having to procure, read and fully understand these standards.

The idea for this book came from countless inquiries - by telephone, in person or by email over many years from people who were confronted with the topic and were looking for support.

Recurring questions about the topic of measuring equipment management and calibration such as "why do you have to ...", "where does it stand ...", "can I also ..." should be answered in a compact manner in this book. It was considered important to work closely on the normative references and to give the necessary references there.

This book is partly identical to the book "Measuring equipment management and calibration" by the same author, BoD, ISBN 9783750434189, but the parts "Structure of measuring equipment management" and "Tips and tricks" are omitted.

Calibration compendium

Definitions

All general definitions refer to:

Burghart Brinkmann
Internationales Wörterbuch der Metrologie
Grundlegende und allgemeine Begriffe und zugeordnete Benennungen (VIM)
German –english edition
ISO/IEC-Leitfaden 99:2007
Korrigierte Fassung 2012

This publication is the reference for all metrological terms in this book.

Calibration compendium

Measurement technology in everyday life

Ensuring product quality is of increasing importance for every company, particularly with regard to the need to maintain or consolidate its economic position in the market.

High quality requirements for a product nowadays mean that an adequate quality management system must be in place (keyword "product liability").

These findings are not new - the modern technology and the possibilities in mechanical production as well as the possibilities of electronic measurement data acquisition and utilization have replaced previous manufacturing processes. There is no longer a "fit" or a "thumbs value".

The pressure to act economically and modern manufacturing technology lead to process-based processes.

In terms of the core factors, business processes and technical or manufacturing processes hardly differ. In order to reduce blurring, it should be noted that the further considerations and explanations in relation to services only relate to technical processes..

Calibration compendium

Metrology – basic categorization

From a holistic perspective, measurement technology is referred to as metrology. Metrology is the study of dimensions and systems of measurement. In the 3rd edition of the VIM in 2007, metrology is defined as "science of measurement and its application".

To meet the expectations and requirements described above, the large area of metrology can be divided into three basic categories:

- Legal metrology
- scientific metrology
- industrial metrology

Calibration compendium

Legal metrology

The basic tasks and goals of legal metrology are enshrined in the calibration law, which also contains the European requirements applicable in Germany. The purpose of this law is to protect consumers in the acquisition of measurable goods and services and, in the interest of fair trade, to create the conditions for correct measurement in business transactions, to ensure measurement security in health protection, occupational safety and environmental protection and in similar areas of public interest and to increase confidence in official measurements.

In the "Organization Internationale de Métrologie Légale" (OIML), representatives from almost 100 countries are working on uniform construction and testing regulations for all measuring devices. In the certification system of the OIML, the certificates issued by the member states certify that a certain type of measuring device complies with the recommendations of the OIML. This means that a type tested and approved in one country can be approved in another without repeating the test.

In Germany there is a national group, the Working Group on Measurement and Verification (AGME).

Calibration compendium

AGME is the coordinating body for the calibration supervisory authorities. It includes the heads of the calibration supervisory authorities of the federal states and, as a guest, a representative of the Physikalisch-Technische Bundesanstalt (PTB). The chair changes every two years.

An office has been set up to create a contact person that does not change every two years for business associations and partners in the other member States.

The working group implements the decisions taken in the national committees for enforcement by the verification offices and the state-recognized test centers. The technical, organizational and legal questions relevant for a uniform implementation in practice are coordinated. Details can be found in the AGME rules of procedure.

Calibration compendium

Scientific metrology

Scientific metrology is not a specialty dedicated to universities or research institutions only.

It is generally known that the national metrology institutes such as the Physikalisch-Technische Bundesanstalt for Germany or NIST in the US, are the "guardians" for the national standards. What exactly is being done there is less known - the results of scientific metrology have a direct - albeit perhaps delayed - influence on everyday measurement technology.

The three main areas of development and activity are:

1. Definition of internationally accepted and recognized units - the kilogram for example

2. The establishment and maintenance of a national and international traceability of every physical measurement variable and connection option from simple measuring equipment to transfer and usage standards to the national standard.

Calibration compendium

3. Realization of the representation of the individual parameters by stable, worldwide repeatable techniques (see also "the international unit system SI").

For example, IPTS68, a (scientific) definition of the temperature scale, has been regarded as an international temperature reference for a long time. In 1990 – almost unnoticed by the public, but consistently implemented in the following years - a new temperature reference was implemented with the ITS90. All calibrated temperature measuring devices - also the clinical thermometer for just under 5 euros from the pharmacy e.g. – are calibrated based on this technical reference nowadays.

Calibration compendium

Industrial metrology

Industrial metrology or measurement technology is the measurement technology that is used every day in development and production and that is largely responsible for the quality of everyday products, but also for safety in many areas of everyday life.

In industrial metrology, a distinction must be made between consumer technology and professional industrial measurement technology.

There is a very complex measuring device landscape in industrial measurement technology:
In the simple case, a craft company uses a number of measuring and testing devices with which service work is carried out. Examples: a radio or television repair shop uses multimeters, oscilloscopes or analyzers; an auto repair shop has multimeters; a brake test bench or torque wrenches.

In manufacturing companies such as the automotive industry, numerous systemic or general, commercially available measuring devices and systems are used continuously.

Calibration compendium

In the automotive industry there are a number of metrological focuses: e.g. fastening technology, dimensional measurement technology, press-fit technology and the large area of electronic measurement technology.

Electronics can be found everywhere in a modern motor vehicle. An unmanageable variety of measuring devices and system analyzers are available to the manufacturers.

But there are also ubiquitous test and measurement systems in screw connection technology: practically every screw in a motor vehicle is subject to its own, strictly observable guidelines. While obvious screw connections such as on wheel nuts or cylinder heads are generally known, it is hardly noticed that e.g. Screw connections for the restraint system are also a problem area to be addressed - here, almost unnoticed by the public, there has already been a recall campaign for at least one manufacturer due to faulty or bad screw connections.

This huge need for measurement technology led to another branch of industry: the production, maintenance and ultimately calibration of systems for manufacturing.

The above systems e.g. in the automotive industry are used for bolt analysis or torque monitoring.

Calibration compendium

Today quality management systems are used in all manufacturing processes. As a rule, these systems define processes by which production is carried out. However, a process is not just a simple requirement to do something according to defined rules - a process is also essentially characterized by control elements with which the process can be kept "on track" and corrected. In technical processes, one or more measured values are always the basis for a qualitative assessment and the basis for decision-making for an "OK" or intervention in the adjustment process.

Calibration compendium

International SI system

All measurements and measurements can be traced back to the only seven basic parameters of the international SI system (System International).

These basic units are
- Length
- Mass
- Time
- Electric current
- Thermodynamic Temperature
- Amount of substance
- Luminous intensity

While most of the measured variables are also known in everyday life, the amount of substance and light intensity can be assigned less.

But it is precisely here that the importance of measurement technology and the influences on everyday handling can be clarified: If you previously bought lamps, they were incandescent or halogen lamps. The specification of the power consumption ("wattage") was equated in the minds of the buyers with an expected brightness.

This assignment no longer works until the marketability of modern LED lights. The lamp manufacturers now rather indicate the light intensity of their product - mostly in lumens.

Calibration compendium

A dimension that was hard to imagine or grasp for many people is slowly becoming "face", you can imagine something under a lumen and compare products with each other.

All physical quantities are organized in a system of dimensions. Each of the seven basic sizes of the SI has its own dimensions, which are symbolically represented by a single capital letter in an upright (not italic) basic font without serifs.

Calibration compendium

Defining natural constants

Since May 20, 2019, seven natural constants have formed the basis of the new international SI system and thus the basis for internationally comparable trade fairs:

- unperturbed ground-state hyperfine transition frequency of the caesium 133Cs-atom
  $\Delta v = 9\ 192\ 631\ 770\ s^{-1}$
- Speed of light in vacuum
  $c = 299\ 792\ 458\ m\ s^{-1}$
- Planck constant
  $h = 6{,}626\ 070\ 15 \cdot 10^{-34}\ J\ s)$
- Elementary charge
  $e = 1{,}602\ 176\ 634 \cdot 10^{-19}\ C\ (C = A\ s)$
- Boltzmann constant
  $k = 1{,}380\ 649 \cdot 10^{-23}\ J\ K^{-1}$
- Avogadro constant
  $N_A = 6{,}022\ 140\ 76 \cdot 10^{23}\ mol^{-1}$
- Luminous efficavy
  $K_{cd} = 683\ cd\ sr\ W^{-1}$

The goal for a long time was not to use (physical) artifacts to define unity, but a natural constant.

This goal has now been achieved for the first time.

Calibration compendium

SI-Units

Transfered into the SI basic units the actual definition is:

Second (s)
1 s = $9\,192\,631\,770/\Delta v$

Metre (m)
1 m = $(c/299\,792\,458)$ s
= $30{,}663\,318\ldots\,c/\Delta v$

Kilogram (kg)
1 kg = $(h/6{,}626\,070\,15 \cdot 10^{-34})$ m^{-2} s
= $1{,}475\,521\ldots \cdot 10^{40}$ h $\Delta v/c^2$

Ampere (A)
1A = $e/(1{,}602\,176\,634 \cdot 10^{-19})$ s^{-1}
= $6{,}789\,686\ldots \cdot 10^8$ Δv e

Kelvin (K)
1K = $(1{,}380\,649 \cdot 10^{-23}/k)$ kg m^2 s^{-2}
= $2{,}266\,665\ldots\,\Delta v\,h/k$

Mole (mol)
1 mol = $6{,}022\,140\,76 \cdot 10^{23}/N_A$

Candela (cd)
1 cd = $(K_{cd}/683)$ kg m^2 s^{-3} sr^{-1}
= $2{,}614\,830\ldots \cdot 10^{10}$ $(\Delta v)^2$ h K_{cd}

Calibration compendium

The International System of Units (SI) is supported by almost 100 countries. The SI is now undergoing a fundamental refresher so that it can calmly face all scientific and technical challenges of the 21st century.
Natural constants such as the speed of light or the charge of the electron will provide the units with the best possible basis for definition.
As early as 1900, when Max Planck developed his radiation law, "constants" and the idea of "natural units of measure" came into play, valid for "all times and for all, including extraterrestrial and extra-human cultures".

All units other than those shown above are derived units. They can be expressed by the basic quantities using physical equations. Their dimensions are represented as a product of powers of the dimensions of the basic quantities using the equations that link the derived quantities to the basic quantities.

These basic units, which are kept, maintained and passed on by the Physikalisch-Technische Bundesanstalt, are of course not an end in themselves. They are the basis for legal metrology in the Federal Republic of Germany.
All measurements that are made relevant to quality or quantity in goods traffic, healthcare, critical technical applications and other areas are traced back to these basic units at PTB through various statutory or quality-related requirements.

Calibration compendium

The characters used for the base sizes and the characters used to indicate their dimension are as follows:

Base parameter	SI	Symbol	Dimensional symbol
Länge	Metre	l, x, r, etc.	L
Masse	Kilogram	m	M
Zeit, Dauer	Second	t	T
Elektrischer Strom	Ampere	I, i	I
Thermodynamische Temperatur	Kelvin	T	Θ
Stoffmenge	Mole	n	N
Lichtstärke	Candela	Iv	J

Table 1: SI-units

Calibration compendium

National metrology institutes

The national metrological institutes are generally responsible for the national implementation of the SI.

Examples:

- Federal Republic Of Germany:
  Physikalisch-Technische Bundesanstalt (PTB) (in former GDR: Amt für Standardisierung, Meßwesen und Warenprüfung ASMW),
- Switzerland:
  Eidgenössisches Institut für Metrologie (METAS),
- Austria:
  Bundesamt für Eich- und Vermessungswesen (BEV),
- Great Britain:
  National Physical Laboratory (NPL) und
- USA:
  National Institute of Standards and Technology (NIST).

An application obligation of the SI arises only through laws or case law of individual states.

Laws governing the introduction of the SI came into force in 1970 in the Federal Republic of Germany (Law on Units and Time), 1973 in Austria (Law on Dimensions and Calibration), 1974 in the GDR and 1978 in Switzerland; In 1978 all transitional arrangements regarding non-SI units were completed.

Calibration compendium

In the EU, the use of units in the area of legal metrology has been largely standardized, among other things, by EC Directive 80/181 / EEC.

In the European Union, Switzerland and most other countries, the use of the SI in official or business transactions is required by law. the use of additional units in the EU was permitted indefinitely with Directive 2009/3 / EC (previous guidelines only allowed this until December 31, 2009).

This was enforced in order not to hinder exports of goods to third countries.

Calibration compendium

Accreditation

The term "accredit" comes from Latin and means "believe".

In view of globalization, it is important to set a standard and a guideline for the requirements for the quality of goods and services so that comparability is possible at all. Only then can inferior products be identified and complained about.

The basic measure of all things in the literal sense for all metrological matters is the safekeeping and transmission of the parameters of the national standards at the Physikalisch Technische Bundesanstalt PTB in Braunschweig.
However, the PTB cannot carry out the task of passing on the measurement technology to the end consumer - "intermediate" positions (of the economy) are required. Since these now perform tasks that have to meet the highest technical and administrative requirements, but are of course also subject to economic interests, a conflict could arise.

Calibration compendium

In order to create clear rules for creating the calibration - depending on the parameter - of comparable quality and minimum requirements, calibration laboratories are "approved", if they are met and if specified requirements are met.

This accreditation is not exclusively based on a paper layer; An on-site accreditation audit is also a prerequisite and is always carried out as a first and later as a follow-up audit.

This audit and the subsequent assessments ensure that the products, processes, services or systems checked are reliable in terms of quality and safety, that they correspond to a minimum technical level and that they comply with the requirements of the relevant standards, guidelines and laws. This is why these objective confirmations are also referred to as conformity assessments.

Calibration compendium

DKD

The abbreviation "DKD" stands for "German calibration service". Until the beginning of 2010, the DKD was the body that accredited calibration laboratories from the private sector and from authorities and the military through the German Accreditation Council DAR.

The DKD has been an association of calibration laboratories in industrial companies, test institutions, technical authorities and was intended to ensure a nationwide metrological presence in Germany. In order to become a member of the DKD, the calibration laboratory had to demonstrate its personnel and measurement expertise as part of an accreditation. It was then authorized to carry out calibrations for the accredited measured variables and measuring ranges and to document them in a DKD calibration certificate. Due to the accreditation and the associated monitoring of the capabilities of the calibration laboratory, this DKD calibration certificate was used as proof of traceability in accordance with ISO 17025 and ISO 9001.

Calibration compendium

The accreditation system was changed in Germany on January 1, 2010 on the basis of Regulation (EC) No. 765/2008. The accreditation body of the German Calibration Service (DKD) was subsequently transferred to the German Accreditation Body GmbH (DAkkS) on December 17, 2009.

However, the DKD specialist committees will continue to operate as technical bodies under the patronage of PTB.

Calibration compendium

DAkkS

DAkkS is the abbreviation for "German accreditation body". This body is the national accreditation body of the Federal Republic of Germany. It acts according to the Accreditation Body Act (AkkStelleG) as the sole service provider for accreditation in Germany.

The DAkkS is a non-profitable organization. It is a llc, whose partners are the Federal Republic of Germany, the federal states of Bavaria, Hamburg, Lower Saxony, North Rhine-Westphalia and Saxony-Anhalt and through the Federal Association of German Industry e. V. (BDI) represent the German economy.

Among other things, the DAkkS accredits calibration laboratories, i.e. After a (strict) check of compliance with specified requirements, a calibration laboratory is "approved". It has demonstrated that it works in accordance with the DAkkS guidelines and may use the DAkkS logo in its calibration certificates.

Calibration compendium

Accredited calibration laboratory

If one speaks of an accredited calibration laboratory in Germany, this always means an accreditation by the DAkkS on the basis of DIN EN ISO / IEC 17025, "General requirements for the competence of testing and calibration laboratories".

DIN EN ISO / IEC 17025: 2018 says - new and in contrast to the previous versions:

This document has been developed with the objective of promoting confidence in the operation of laboratories. This document contains requirements for laboratories to enable them to demonstrate they operate competently, and are able to generate valid results. Laboratories that conform to this document will also operate generally in accordance with the principles of ISO 9001.

If the message "Accreditation according to 17025" did not necessarily have something in common with "Certification according to DIN EN ISO 9001" in the earlier document, the standardization committees reacted and recognized the reality

- That the vast majority of accredited calibration laboratories also have ISO 9001 certification
- That there are numerous intersections in terms of content - here: the general QM shares - in the two quality management manuals

Calibration compendium

- that these two segments are not contradictory, but rather complement each other, and the dealings used up to now mean double work and double burden for all parties

Accordingly, the new edition of DIN EN 17025: 2018 can / should create a common, general QM component and use it for certification and accreditation.

Nevertheless, when looking for a qualified calibration laboratory, the focus should not be on certification according to ISO 9001, but on accreditation according to DIN EN ISO / IEC 17025.

While ISO 9001 places fundamental demands on quality management, DIN EN ISO / IEC 17025 is specially tailored to calibration laboratories.

Calibration compendium

In this regulation, the technical requirements for a calibration laboratory alone include a catalog of measures for:

- The staff
- The premises and environmental conditions
- The test and calibration procedures and their validation
- The selection of procedures
- Procedures developed by the laboratory
- Procedures not specified in normative documents
- Validation of procedures
- Estimation of measurement uncertainty
- The control of data
- The facilities
- Metrological feedback
- Special requirements
- Reference standards and reference materials
- Sampling
- Handling of test and calibration objects
- Ensuring the quality of test and calibration results

A calibration laboratory that has been accredited (with the corresponding effort) according to DIN EN ISO / IEC 17025 can always be selected as a trustworthy calibration center for the accredited parameters.

Calibration compendium

Calibrations carried out in an accredited laboratory stand for the reliability of the measurement results (including uncertainties) and are essential for certification according to EN ISO 9001, which requires traceability to national standards for the measuring equipment used.

However, offered or performed calibrations outside the scope of calibration can also be trusted as factory calibrations - a laboratory that does the work described above will not work outside of its quality management system - however, no one can guarantee this. So there is always a decision based on requirements and data, whether a factory or DAkkS calibration is required.

However, calibration services without any certification or accreditation are not advisable - a "fits" calibration has no (metrological) value.

Despite the delimitation to ISO 9001 described above, an accredited calibration laboratory generally also has ISO 9001 certification.

Calibration compendium

This is due to the equally aligned requirements for a process-oriented implementation. These are, however, generally accepted in ISO 9001 and fit both service providers and manufacturing companies.
For this reason, DIN EN ISO 10012, "Measurement management systems. Requirements for measuring processes and measuring equipment "created.

This regulation supplements the above-mentioned DIN EN ISO / IEC 17025 and ISO 9001 and provides the calibration laboratories with tools and guidelines for implementing a calibration with high quality standards and comparable quality.

Calibration compendium

Normative references

In December 1989, the Product Liability Act (Law on Liability for Defective Products - ProdHaftG) came into force in Germany. It regulates a manufacturer's liability for defective products. The law aims to compel the manufacturer with the greatest care in the manufacture of products in order to minimize or exclude the dangers that may arise from a product and thus to protect the consumer. While the law regulates the liability conditions, supplementary regulations and standards provide enforcement and implementation requirements.

One of these requirements is the ISO 9000 series of standards, according to which many (not just manufacturing) companies are certified today. It specifies requirements for quality management systems, which should serve as tools to achieve defined quality standards and thus product safety and the protection of the end user.

Certification according to ISO 9000 is proof that the due diligence required by the Product Liability Act has been met by setting up suitable processes and establishing a holistic quality management system.

Calibration compendium

However, there are a large number of QM-relevant standards or standards / regulations with calibration reference.

The most important references are to be mentioned below. The list is not exhaustive.

- DKD 4 Traceability of measuring and testing equipment to national standards
- International vocabulary of metrology. Basic and general concepts and associated terms (VIM) – German – English version ISO/IEC-Guide 99:2007 (Beuth Wissen) Taschenbuch – 7. August 2012 von DIN e.V. (Herausgeber), Burghart Brinkmann (Autor)
- DIN EN ISO/IEC 17000:2005-03: Conformity assessment - terms and general basics
- DAkkS-DKD-5: Instructions for creating a calibration certificate (Section A, Item 2)
- DIN EN ISO 10012:2003: Measurement management systems - requirements for measurement processes and measuring equipment
- ILAC-G8:03/2009: Guidelines on the Reporting of Compliance with Specification
- UKAS M3003: The Expression of Uncertainty and Confidence in Measurement (Edition 2, 2007), Appendix M Assessment of Compliance with Specification

Calibration compendium

The three most current standards should now be addressed in part, relevant passages shall be discussed:

- DIN EN ISO 9001:2015
  Qualitätsmanagementsysteme – Anforderungen
- IATF 16949
  Quality management system requirements for automotive production and relevant service parts organizations
  (BMW, Chrysler, Daimler, Fiat, Ford, General Motors, PSA, Renault, VW)
- DIN EN ISO 17025:2018
  General requirements for the competence of testing and calibration laboratories

ISO DIN EN ISO 9001:2015

For the field of measuring and testing equipment, an essential section of ISO 9001: 2015 is section 7.1.5 ff. It is the binding basis for the construction and operation of a measuring equipment management system:

ISO DIN EN ISO 9001:2015, Ziff 7.1.5 ff

The organization shall determine and provide the resources needed to ensure valid and reliable results when monitoring or measuring is used to verify the conformity of products and services to requirements.

The organization shall ensure that the resources provided:

a) are suitable for the specific type of monitoring and measurement activities being undertaken;

b) are maintained to ensure their continuing fitness for their purpose.

The organization shall retain appropriate documented information as evidence of fitness for purpose of the monitoring and measurement resources.

Damit werden feste Rahmen für den Einsatz und die Nutzung von Mess- und Prüfgeräten in einem (zertifizierten) Betrieb vorgegeben.

Calibration compendium

The following paragraph makes it very specific:

„7.1.5.2 Measurement traceability
When measurement traceability is a requirement,... „

In the introductory sentence above the norm allows an exception with the word "when" - obviously not every measuring and testing device is affected by the measures listed below. This means that when setting up a measuring equipment management system, one should deal with the measuring devices in operation not only with regard to their function but also with regard to their application.

... or is considered by the organization to be an essential part of providing confidence in the validity of measurement results, measuring equipment shall be:

> a. *calibrated or verified, or both, at specified intervals, or prior to use, against measurement standards traceable to international or national measurement standards; when no such standards exist, the basis used for calibration or verification shall be retained as documented information;*

In addition to the technically-oriented requirement to maintain and use precise measuring and testing devices, this sentence also includes a sustainability-oriented approach:

Calibration compendium

It is therefore not sufficient to procure a measuring and testing device calibrated or to have it calibrated once, but this measure must be carried out at intervals or repeated depending on the application.

In this (current) version of ISO 9001, a traceable calibration is even required. This is often interpreted as meaning that every measuring device must be traceably calibrated. The exact wording is:

„... *against measurement standards traceable to international or national measurement standards, ...*"

This allows the use of a factory standard, on which the used measuring equipment is calibrated. Example: a calibration system for torque wrenches is traceable - in Germany by DAkkS calibration - calibrated and is available in operation (= "the organization"). Depending on the application, the operational torque wrench can be calibrated as a manufacturers calibration.

In the following text, the representative or user of the measuring device is specifically encouraged to visually mark his devices and to take protective measures:

Calibration compendium

The measuring equipment shall be
...*identified in order to determine their status*

Examples can be:

An own provided calibration label:

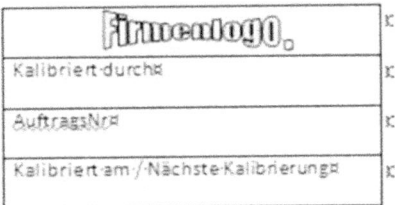

Or, in the event that a device is not calibrated (and therefore may not be used in a quality-relevant manner), a sticker such as:

„not calibrated"

An auditor - much more important but in the everyday use of the user - recognizes the status of the measuring device at a glance.

Calibration compendium

It can be discussed whether the displayed decision (e.g. "not calibrated") is correct. However, this takes an audit to another level. It is important: the auditor recognizes that all measuring devices have been taken into account and decisions regarding their calibration status have been made.

The standard also specifies:

The measuring equipment must ...

> b. *safeguarded from adjustments, damage or deterioration that would invalidate the calibration status and subsequent measurement results.*

Here too, the measuring device holder is in direct duty:

He must take suitable measures to prevent (at least) unintentional manipulation of the measuring device.

It also says:

> *The organization shall determine if the validity of previous measurement results has been adversely affected when measuring equipment is found to be unfit for its intended purpose, and shall take appropriate action as necessary.*

Calibration compendium

This requests a look into the (recent) past regarding the use of the measuring device:

- where / on which object has been measured with the measuring equipment

- does this have any quality-relevant effects?

- Does this even have security-relevant effects (aviation or automotive security!)?

- What must be done if one of the last two questions is answered in the affirmative: recall? Customer Information? New production line?
  ...

Calibration compendium

IATF 16949

The IATF (International Automotive Task Force) is a "purpose-specific" working group, which is made up of representatives of mostly North American and European automobile manufacturers and automobile associations and deals with the harmonization (standardization) of standards (norms) for improving product quality for automotive customers.

The following automobile manufacturers are full members of the IATF:

- BMW Group
- Daimler AG
- Fiat Auto
- Ford
- General Motors
- PSA (Citroën, DS, Opel, Peugeot, Vauxhall)
- Renault
- Volkswagen AG

as well as the following national associations:

- AIAG (North America)
- ANFIA (Italy)
- FIEV (France)
- SMMT (Great Britain)
- VDA (Germany)

Calibration compendium

With its regulations, the IATF has great power and clearly influences the calibration market.

The OEMs (Original Equipment Manufacturer) that are members of the IATF require their suppliers to be certified according to IATF 16949.

IATF 16949: 2016 is decisive for the area of quality assurance and especially for the requirements for the calibration of measuring and testing devices.

Episode:

Factory calibrations are hardly in demand anymore, measuring instrument holders in the automotive industry or their suppliers nowadays almost exclusively require traceable calibrations - DAkkS calibrations in Germany.

This is not always required in the regulation - the IATF 16949 allows a wide range of possibilities - but which are often not known or are overlooked prematurely:

The IATF 16949 (based on point 7.1.5.3.2), which is based on EN ISO 9001: 2015, directly relates to the quoted DIN EN ISO 9001: 2015-11:

Calibration compendium

An external calibration service provider can be commissioned if no own laboratory (note: at the measuring device holder) is available for calibrations and the external laboratory has accreditation according to DIN EN ISO / IEC 17025 and can provide proof of its suitability.
Proof of suitability should meet the standards of DIN EN ISO / IEC 17025.

Other permitted calibration centers, such as device manufacturers, are approved according to IATF 16949 in cases where it is difficult to find a suitable qualified (accredited) laboratory for every measurement parameter.
System manufacturers and operators of DAkkS-accredited laboratories generally meet these requirements.

Since there are always questions about the implementation of IATF 16949, the IATF (International Automotive Task Force) has written and released a series of "Sanctioned Interpretations".

Calibration compendium

When asked whether an accredited body must always be used as a calibration laboratory, the "Sanctioned Interpretation Nov. 2018" makes the following stipulations to Section 7 / IATF 16949 reference 7.1.5.3.2:

- *"The manufacturers of measuring and testing equipment also develop the methodology for compliance and adjustment of their measuring or testing equipment as part of their development and manufacture to ensure that the respective calibration requirements are met. For this reason, the manufacturer of the measuring and testing equipment is fundamentally qualified to calibrate the equipment he has developed and manufactured.*

- *However, the organization must obtain the customer's approval before using the manufacturer of the measuring and testing equipment for calibration services.*

It is therefore not always necessary to commission a laboratory accredited for the respective parameter with the calibration. The decisive factor - this applies generally to all calibrations - is to choose a laboratory with appropriate qualifications and its evidence:

Calibration compendium

Ideally, this includes ISO EN 17025 accreditation in one parameter (this gives the laboratory a "framework" for demonstrating its competence.

In any case, care must be taken that the definition of the term "calibration" is understood and implemented: a look at the calibration certificate issued by a laboratory helps here; are measurement uncertainties indicated - or only comparative values? Comp. "Calibration certificate".

Calibration compendium

DIN ISO/IEC 17025:2018

The DIN ISO / IEC 17025: 2018 standard is THE standard according to which a calibration laboratory can be accredited - i.e. approved - by the German Accreditation Service (DAkkS), provided that the laboratory fulfills a whole range of requirements. It

- It regulates the competence, impartiality and consistent functioning of laboratories of all sizes, as well as their customers as well as accreditation and certification bodies, which with the help of DIN EN ISO / IEC 17025 can understand or demonstrate the competence of laboratories.

- is part of the ISO / IEC 17000 series of standards for conformity assessment.

- contains, in addition to specific requirements for laboratories, general text modules that are repeated throughout the ISO / IEC 17000 series of standards. This applies, for example, to the requirements for impartiality, confidentiality, the handling of complaints and management.

Calibration compendium

- underwent a comprehensive revision, application area was streamlined and the entire structure was revised.

- now includes: consideration of risks and opportunities; Impartiality and confidentiality were emphasized; Competence of the staff was emphasized.

Laboratories that meet this draft standard will generally also work in accordance with the principles of ISO 9001.

However, this standard is also important for the user of calibration services: it contains specifications for the calibration laboratory that must be implemented and received directly by the user of the measuring device and that can be requested. There are guidelines for the content of calibration certificates and the declaration of conformity that may be contained therein.

The section "Calibration Certificate" is therefore dedicated to this part.

Calibration compendium

DIN ISO/IEC 10012:2004-03

The standard DIN ISO / IEC 10012: 2014 forms the normative basis for measurement management systems.

A measuring equipment management system ("measurement management system") is also intended to ensure that measuring equipment and processes are suitable for the intended use, that incorrect measurement results may be recognized early and that the risk and the effects can be managed.

In the following section, sections are shown that directly affect measuring device holders / users. This group of people is addressed in the foreword to the standard:

References to this International Standard can be made
- *by a customer when specifying products required,*
- *by a supplier when specifying products offered,*
- *by legislative or regulatory bodies, and*
- *in assessment and audit of measurement management systems.*

Calibration compendium

However, this standard gives some striking specifications that also directly affect the measuring equipment holder / user:

7.1 Metrological confirmation
7.1.1 General
> *Metrological confirmation (...) shall be designed and implemented to ensure that the metrological characteristics of the measuring equipment satisfy the metrological requirements for the measurement process. Metrological confirmation comprises measuring equipment calibration and measuring equipment verification.*

The clear obligation to carry out (regular) calibrations is expressed here. The meaning of the specification of measurement uncertainties, without which one cannot speak of a technically correct and complete calibration, is emphasized in the following text as a means of evaluation.

Calibration compendium

Specifications are also made for the determination of the calibration intervals:

7.1.2 Intervals between metrological confirmation
The methods used to determine or change the intervals between metrological confirmation shall be described in documented procedures.
These intervals shall be reviewed and adjusted when necessary to ensure continuous compliance with the specified metrological requirements.
...

Each time nonconforming measuring equipment is repaired, adjusted or modified, the interval for its metrological confirmation shall be reviewed.

Compare also the chapter "Determination and adjustment of calibration intervals" in the back part of this book.

Calibration compendium

The Commercial Measurement System

NIST defines in its NIST handbook 115 / 2011 the commercial measurement system as following:

Many commercial transactions are based on weight, volume, length, or count of products bought and sold. Packaged goods are purchased at the supermarket, people buy delicatessen items over price computing scales, gasoline and diesel fuel are purchased through pumps (retail motor fuel dispensers), gasoline and diesel fuel must meet prescribed quality or octane standards, scanners are used at checkout stands in retail stores to look up prices of products identified by bar codes, farmers sell grain, produce, and livestock over scales, grain prices are adjusted up or down based upon quality measurements, and landfills charge fees based upon the weight of the trash delivered. The structure within which transactions among businesses and with the general public are conducted is called the commercial measurement system.

Weights and measures activities are pervasive within the United States. It is estimated that U.S. weights and measures regulations impact roughly half of the U.S. gross domestic product. The success of the commercial measurement system can be judged by the ease with which transactions are executed, the level of confidence that buyers and sellers have, and the accuracy with which these transactions are performed.

Calibration compendium

In a well-functioning commercial measurement system, effective laws and regulations are in place to ensure an orderly marketplace. The laws and regulations should provide consumer protection by preventing deceptive and misleading practices, but should not be overly burdensome to businesses. They should also foster fair competition among companies in the many different facets of the commercial measurement system. Finally, the laws, regulations, and technical standards must be sufficiently flexible to adjust to new technology and marketing practices. Determining the correct balance of these many factors is a major and ongoing challenge to the weights and measures community

Calibration compendium

Eichung (Germany)

In Germany, legal metrology issues are performed by „Eichämter". These institutes perform calibrations similar to the US wherever the customer is on focus like gas stations masses and weights for selling goods etc. But, these institutes do not line up in the line of accredited and traceable laboratories:

Die Deutsche Akkreditierungsstelle DAkkS teilt mit einem Informationsschreiben zum „Merkblatt zur metrologischen Rückführung im Rahmen von Akkreditierungsverfahren (71 SD 0 005 Revision 1.4) vom Februar 2016 mit:

„Revision 1.4 is a limitation of the previous options, additional options will not be introduced.
The restriction concerns the revision of the conditions for repatriation via so-called "results reports without accreditation symbol" in point 6 of the information sheet. Point 6.c) "Result reports without accreditation symbol, issued by German verification authorities" was deleted without replacement. This means that results reports from German calibration authorities are treated in exactly the same way as other results reports without an accreditation symbol that have been issued by non-accredited bodies.

The recognition of verification certificates as evidence of metrological traceability is therefore only possible in individual cases. The necessary assessments of the bodies that issue results reports (i.e. return reports) are only

Calibration compendium

carried out by the DAkkS. Results reports from German calibration authorities can e.g. B. Considered as return evidence if there are demonstrably no other bodies that can perform these calibrations.

A measuring device with an "Eichamt" calibration certificate is therefore not considered to be traceable calibrated!

Calibration compendium

Why calibrate?

All measuring and testing devices are exposed to influences during use and also during storage, which can permanently change their metrological properties. Calibration determines whether measuring and testing devices have the required accuracy.

Definition calibration:
The international vocabulary of metrology defines calibrations as:

*„operation that, under specified conditions, in a first step, establishes a relation between the **quantity values** with **measurement uncertainties** provided by **measurement standards** and corresponding **indications** with associated measurement uncertainties and, in a second step, uses this information to establish a relation for obtaining a **measurement result** from an indication*

*NOTE 1 A calibration may be expressed by a statement, calibration function, **calibration diagram**, **calibration curve**, or calibration table. In some cases, it may consist of an additive or multiplicative **correction** of the indication with associated measurement uncertainty.*

Calibration compendium

NOTE 2 *Calibration should not be confused with **adjustment of a measuring system**, often mistakenly called "self-calibration", nor with **verification** of calibration.*

NOTE 3 *Often, the first step alone in the above definition is perceived as being calibration.*

Calibration is therefore a measurement process for reliably reproducible determination and documentation of the deviation of a measuring device or a material measure to another device or another material measure (normal).

According to the definition of the VIM by JCGM 2008, a second step is mandatory to define the calibration: Taking the determined deviation into account when using the measuring device to correct the readings. These definitions, which are formulated very theoretically, should be explained below:

Calibration compendium

Each measuring device / device is characterized by three basic parameters; these are parameters of each calibration:

- ✓ Precision:
  How accurate is the displayed value?

- ✓ Repeatability:
  Can this value be repeated x times for x measurements?

- ✓ Linearity:
  On a scale of 0 - 100 of any measurement variable (ampere, N m, volt, kg or similar) - does the displayed value always have the same storage?
  Example: If 18 - 28 - 38 - 48 etc. is displayed, the measuring device has a storage of - 2. This storage is easy to correct: either by means of a correction table or by means of a comparison. In today's world, such a correction table will lead to a "correct" display via software (blue example line).

Calibration compendium

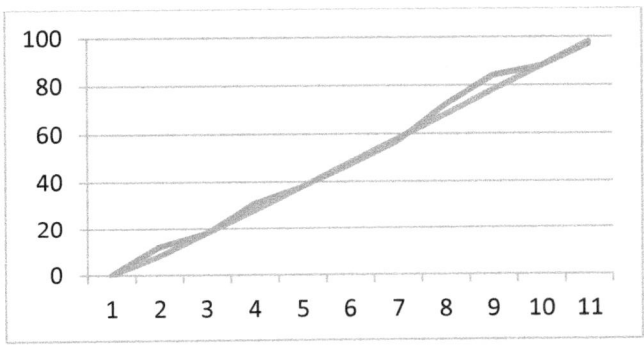

But the meter would display 18 - 31 - 38 - 47. The displayed value "balls" around the true value - the measuring device is non-linear and much more critical to treat - and definitely of poorer quality (red example line).

The characteristics mentioned should be illustrated in the following graphic:

1. The measurement result spreads over a wide range - the measuring device is unusable for reliable measurements.
2. The measurement result is not very precise, but can be repeated in a certain (wide) range - the measuring device is for certain tasks can be used Examples: folding rule , caliper, coarse weights.

Calibration compendium

3. The spread is low (= good repeatability), but has a shelf. The meter is good, but needs a correction (table or software).
4. Good, precise measurement result with high repeatability and small linearity error.

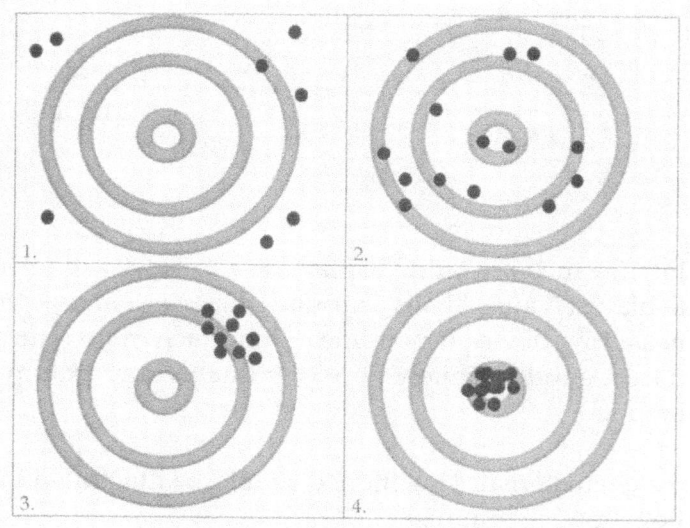

With a calibration, the relationship between the output values of a measuring device or a measuring device or the values represented by a material measure or by a material measure and the associated values of a measured variable determined by standards is determined under predetermined conditions.

Calibration compendium

This means: there is a (precise) standard (which of course must also be calibrated and traceable) and a (poorer) measuring device.
Using a suitable method and under defined conditions (e.g. temperature, air humidity), it is now determined how far the test specimen deviates from the target value.

In this sentence - a matter of course - a number of specifications were mentioned, which are not taken for granted in practice, but are the basis for the quality of a calibration.

So at the end of a calibration it is not a matter of course that the test object is also within its specifications. This is usually the expectation of the measuring device owner - but this does not necessarily have to be met.

A calibration certificate initially only documents the (metrological) properties of the test object found at the time of calibration at an accredited calibration laboratory using a standardized procedure with a validated measurement method and an uncertainty factor. Only if - as possible - a declaration of conformity is required will the customer's expectations be met, provided the measuring device is within the specifications - this does not include "simple calibration".

Calibration compendium

Profile of a measuring device - added value

A measuring and testing device must (can) be trusted - the confidence to believe that it is within its specifications and always measures "exactly" when used properly and professionally.

In a technically oriented environment of highly complex systems, trust - at least in the colloquial sense - is not a factor on which to build and plan.

Calibrations create this trust. Regular calibrations create a data-based and thus reliable basis for such trust.

With a calibration, it can be determined whether the measuring and testing device is within its specifications. If it does not, the device must be adjusted or repaired.

Data on these measures are available to the device holder as a result of this calibration - for example in the form of a calibration certificate.

Value creation has already begun here, which can be sustainable: if the data of several - at least three calibrations - are collected and evaluated, the calibration interval can be monitored and adjusted
and the measuring device can be classified with regard to its reliability.

Calibration compendium

The product "measuring device" gets its metrological profile only after a first and then through subsequent calibrations, which allows statements about the reliability and its precision.

To develop such a profile it is not absolutely necessary, but it is strongly recommended to always use the same calibration service. This ensures the same conditions and procedures for calibration. With "jumping" between different calibration providers for cost reasons e.g., an evaluation is often not possible due to incomparable calibration scopes or inconsistent calibration certificates.

Calibration compendium

Replacement of a measuring device for cost reasons

The argument is gladly taken: "... just throw away the item; a new purchase is cheaper than a calibration ... "
These arguments are often used for cheaper measuring devices such as torque wrenches, multimeters, feeler gauges, dial gauges etc..
Anyone who argues like this did not understand that they are talking about two products:
1. a measuring device (the "hardware") and
2. establishing and depositing the (metrological) properties of this device.

A measuring device often is a mass product - in many applications, however, it cannot be exchanged arbitrarily - it is a product that was manufactured and whose very specific properties - in the case of a measuring device, a material measure - only through a precise first and subsequent measurement (is documented by a calibration).

Only if calibrations have been carried out regularly by an (accredited) calibration center will you get knowledge about
- the reliability
- the precision
- the stability

of the device, can build up a history and make estimates of stability and calibration interval.

Calibration compendium

Change through use

The Physikalisch-Technische Bundesanstalt has known a whole range of measuring devices - own devices but also customer devices - over many years and decades and monitors their development and stability.

This revealed abnormalities that are the basis for further research.

Example gage blocks:

A regular profile can be developed for sets that are regularly presented for the calibration - for example to the Physikalisch-Technische Bundesanstalt. Astonishing insights were made: There are blocks that grow instead of wear out and grow minimally in length!

Such knowledge is the basis for further research and can only be obtained through regular calibrations or follow-up measurements.

Calibration compendium

Traceability

In connection with calibrations and measuring and testing devices, the term "traceability" is mentioned again and again and traceability is required.

Simply explained, traceability means that any measuring device is calibrated with a more precise measuring device (a standard), which in turn has been calibrated to an even more precise standard - this chain must be continued until the most accurate available standard - usually a national standard, e.g. at the Physikalisch-Technische Bundesanstalt - is achieved.

A whole series of characteristics must be documented at each level in order to demonstrate that the national standard has been reached and thus demonstrate traceability.

The paper DAkkS-DKD-4 *"Traceability of measuring and testing equipment to national standards"* states:

"The requirements for quality management systems are set out, for example, in the ISO 9000 series of standards, which is identical to the European standard series EN ISO 9000. The monitoring, calibration and maintenance of measuring and test equipment is an important part of these standards and guarantees that the measurements are carried out correctly during the manufacturing process. For this purpose, all measurement results must be traced back to national standards ".

Calibration compendium

This defines a clear interpretation of ISO 9000. In order to establish the reference to metrological relationships, the term traceability must also be looked up in the international dictionary of metrology:
This reference knows the term as "metrological traceability" and defines it as follows:

*„property of a **measurement result** whereby the result can be related to a reference through a documented unbroken chain of **calibrations**, each contributing to the **measurement uncertainty**"*

This paragraph includes a number of attributes to consider:

Traceability is the "*property of a measurement result ...*"

The following part of the sentence now lists components that can be assigned to a traceability / condition of a traceability:
- Documentation
- Measurement uncertainty
- Unbroken chain of calibrations

Calibration compendium

Documentation
„...the result can be related to a reference through a documented ...,"
Like all elements of a QM system, written proof of all process steps is essential. This applies to the proof of performed calibrations as well as the proof that these calibrations were carried out on a standard, which in turn was calibrated on a calibrated standard.

A measurement result therefore has further properties in addition to a determined measured value. Consistently implemented, these properties also include a relationship to one or more other measurement results and the associated measurement uncertainties. A determined / read measured value does not stand alone and absolutely in the room, but when viewed correctly and comprehensively belongs to a part of a chain or cascade of further measurements, measurement uncertainties and measurement results:

Measurement uncertainty
*„...each contributing to the **measurement uncertainty**,..."*
According to the definition cited above, traceability is not defined exclusively on the measuring device, but rather on the measurement results that are achieved.

Calibration compendium

What components does a measurement result consist of?
- A measurement almost always leads to a numerical value (e.g. read value).
- A unit is assigned to this numerical value (volts, amperes, newton meters, etc.).
- Unfortunately there is always an error - no measurement is "infinitely accurate"

Expressed as a formula:

$$X_w = X \pm F$$

with
X_w as „true" measured value
X as measuered value
F as error.

If you knew this error, it would be easy to convert the measured value into a true measurement result. Unfortunately, the true size of this error is never known - by appropriately limiting as many known or suspected error components as possible (examples: precision / "measuring accuracy" of the measuring device, temperature influences, measuring conditions, measuring methods, etc.), this error can be described and kept as small as possible.

A whole series of attributes can therefore be assigned to a measured numerical value.

Calibration compendium

An attribute is the measurement uncertainty, in which the described error influences are summarized and which falsify the determined numerical value. In summary, the measured value and the calculated measurement uncertainty is a measurement result.

See also detailed explanations in the "Measurement uncertainty" chapter.

Calibration hierarchy

„... *whereby the result can be related to a reference through a documented unbroken chain of* **calibrations**,"

Measuring devices in the own company are tested or even calibrated on anr own test stand, for example. This test stand functions as a factory or service standard.
Of course, this test stand also has to be calibrated at regular intervals. This chain can / must continue.

The next higher level is the connection via an accredited calibration laboratory. Typically, after about three or four steps, the highest available standard - usually a national standard - is reached in the Federal Republic of Germany at the Physikalisch-Technische Bundesanstalt in Braunschweig or at NIST in the United States.

Calibration compendium

This chain of relationships from measuring device - normal to national normal is called the traceability chain in metrological usage.

The dependency of this traceability chain makes the graphic shown clear. The direction of the arrow represents the direction in which the normal results are passed on.

Calibration compendium

The measuring and testing equipment, as they are often found in large numbers, were designated as consumables; e.g. Multimeters, micrometer screws, oscilloscopes, torque wrenches etc.

A usage or factory standard can often be found in your own company: e.g. with a calibration device for torque wrenches or a calibrator for checking or calibrating the company's own multimeter.

To meet the traceability requirements, these standards must be calibrated by an accredited calibration laboratory - they require a DAkkS calibration certificate.
In some large companies there are even factory owned and run calibration laboratories. These laboratories can calibrate the factory standards for the accredited parameters.
The standards of this calibration laboratory are to be calibrated either directly by the PTB or another competent, accredited calibration laboratory.

Calibration compendium

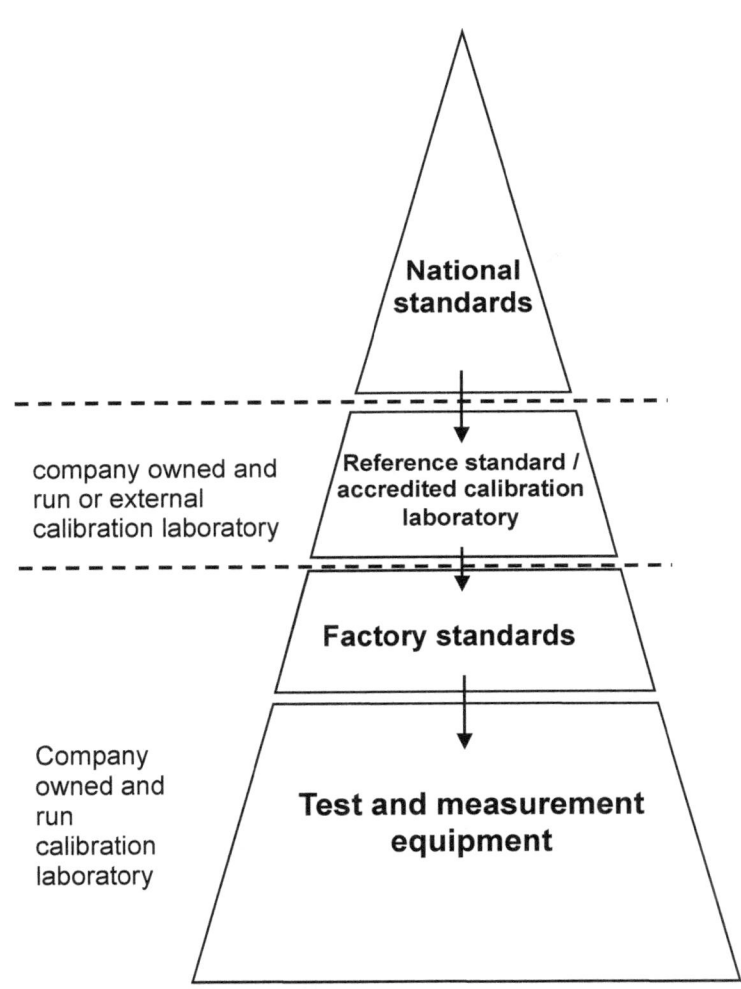

Calibration compendium

Measurement uncertainty

„You always measure wrong, you just have to know how much." [Dave Packard]

As already in the chapter "Traceability" pointed out there are errors that are made with every measurement. Conclusion:

There is always an error in a measurement result - no measurement is "infinitely accurate"

Expressed in a formula again:

$$X_w = X \pm F$$

with

X_w as „true" measured value
X as measured value
F als error.

The measurement uncertainty is always part of a measurement result. It is the "error" inherent in a calibration:
- Even the standard against which it is compared also has an inaccuracy
- The measuring procedure can contain (small) errors
- Uncertainties in the evaluation software
- environmental influences (temperature, humidity, etc.)
- ...

Calibration compendium

These "errors", better known as deviations, can consist of the following components:

Systematic deviations

Systematic deviations are typically known sizes. They can usually be corrected. If this is not possible, these deviations must be added linearly.

Random deviations

Random deviations must be taken into account on the basis of statistical calculations. For this reason, an assessment is also used in this context.

A confidence range is now assigned to these estimates, which is typically 95%: this means that 95 out of 100 measurements lie in this estimated range.

An expansion factor is included in order to raise the measurement uncertainty now determined, which still contains considerable unreliability, to a safe level. This factor is typically assumed to be $k = 2$.

Calibration compendium

This factor comes from Gaussian normal distribution theory and is exactly k = 1.96.

A simple example shall clarify this:

The idea when specifying a measurement result is:

The measurement result MUST be true

Example:
The width of a door should be determined. A class III articulated joint rod ("folding rule") is available. This reads a value of 79.8 centimeters. As shown, this value is only part of the measurement result.

A measurement uncertainty must now be assigned to this value. This measurement uncertainty is determined in the simplest case by setting up a matrix of possible error influences. The individual positions in this list are now assigned to the error. If this error is not known, it can also be estimated (although this worsens the numerical value of the result - i.e. the measurement uncertainty increases), it improves the overall result because it is more true.

Calibration compendium

Example "folding rule":

Error matrix „folding rule"	
Basic accuracy	0,6 mm
Class III	0,4 mm/m
Temperature / Environment	± 2 mm (est.)
Hinges	± 0,5 mm
Handling (Reading, incorrect use)	± 1 mm

First, the error limit of the link rod dimension must be determined:

$$a + b * L$$

with
- L = Size of the length to be measured, rounded up to the nearest full meter
- a, b can be found in the accuracy classes for linear encoders, EC Directive 2004/22 / EC

Note: the error limit is not to be confused with the measurement uncertainty!

In our example, this results in:
$$0,6 \text{ mm} + 0,4 \text{ mm/m} * 1 \text{ m} = 1,0 \text{ mm}$$

The measurement uncertainty can now be calculated: The individual errors are now squared and added. The square root of this sum

Calibration compendium

is the determined measurement uncertainty (geometric addition).
When adding measurement uncertainties, the same dimension of all parts must be observed!

$$u = \sqrt{1^2 + 2^2 + 0{,}5^2 + 1^2 + 1^2} = 2{,}693 \text{ mm}$$

Now the expansion factor is included:
  2,69 mm * 1,96 = 5,27 mm

The "exact" measurement of 79.8 cm therefore has a measurement uncertainty of 5.27 mm.

This seems too high? If you look at the individual components and then consider an interaction under the worst possible circumstances: inaccurate application, inaccurate reading, play in the hinges, extreme e.g. Summer temperatures, among other things, the value is not so unrealistic.

Calibration compendium

The example shown is only intended to help you understand the measurement uncertainty. The calculation for "real" measurement uncertainties is - depending on the parameter - much more complicated, as the following example should show:

Measured variable DC current
The current consumption of the test object shall be measured.

Set up

Carrying out the measurement and evaluating the results
The operating voltage is set on the test object in accordance with the protocol, then read the DC current on the digital multimeter (DMM).

Symbols used

I_{Fluke} :   DMM Reading
ΔI_{Fluke} : Deviatiojn of reading
I_{DUT} :   uut consumed current

Calibration compendium

List of disturbing influences

Parameter X_i	Cause	Value	Weighting factor	Data source
ΔI_{Fluke}	Range / DC-I < 1 A > 1 A	$5 \cdot 10^{-4} + 40$ µA $1 \cdot 10^{-3} + 40$ µA	normal	8440A specs

Factors not considered
none

Derived influencing factors
none

model equation
$$I_{DUT} = \left(I_{Fluke} + \Delta I_{Fluke}\right)$$

Uncertainty budget
It is assumed that all input variables are uncorrelated to one another.

Calibration compendium

Parameter X_i	Cause	Estimated value x_i	Standard-Uncertainty $u(x_i)$	Distribution	Sens.-Koeff. c_i	Effective degree of freedom v_{eff}	Uncertainty $u_i(y)$
I_{Fluke}		0,6 A 0,7 A 1,5 A	-	-	-	-	-
ΔI_{Fluke}	0,6 A 0,7 A 1,5 A	0	$2,9 \cdot 10^{-4}$ $2,8 \cdot 10^{-4}$ $5,2 \cdot 10^{-4}$	normal	1	∞	$2,9 \cdot 10^{-4}$ $2,8 \cdot 10^{-4}$ $5,2 \cdot 10^{-4}$
u	0,6 A 0,7 A 1,5 A	0		----		∞	$2,9 \cdot 10^{-4}$ $2,8 \cdot 10^{-4}$ $5,2 \cdot 10^{-4}$
U	0,6 A 0,7 A 1,5 A			$k = 2$		∞	$6 \cdot 10^{-4}$ $6 \cdot 10^{-4}$ $1 \cdot 10^{-3}$

Specified expanded measurement uncertainties
The expanded measurement uncertainties specified in the table are given.

Calibration compendium

However, there is suitable software for setting up model equations and calculating the expanded measurement uncertainty.

GUM serves as a reference:
GUM is the abbreviation for the ISO / BIPM guide "Guide to the Expression of Uncertainty in Measurement".
It was first published in 1993 and last revised in 2008. The relevant German version is the pre-standard DIN V ENV 13005 (current edition: 1999-06) "Guideline for specifying the uncertainty when measuring".

The GUM was implemented in a "GUM Workbench" software. More information on this at http://www.metrodata.de/.

A free online calculation of measurement uncertainties can e.g. via the homepage of the NIST (National Institute Of Standards), the US equivalent of PTB: http://uncertainty.nist.gov/.

Calibration compendium

Messunsicherheit oder Toleranzangabe

Unfortunately there is always confusion about these two terms. If a measuring device does have a tolerance (example: ± 3%) - what is the measurement uncertainty all about?

When producing a quality product, this product is guaranteed to have certain properties. An example - deliberately not chosen for a measuring device - should clarify:

A door with a width of 90 cm is to be manufactured. A (fictitious) specification requires a manufacturing tolerance of ± 1%.
This means that if this requirement is met, no door leaves the factory that does not have a width between 89.10 and 90.90 cm:

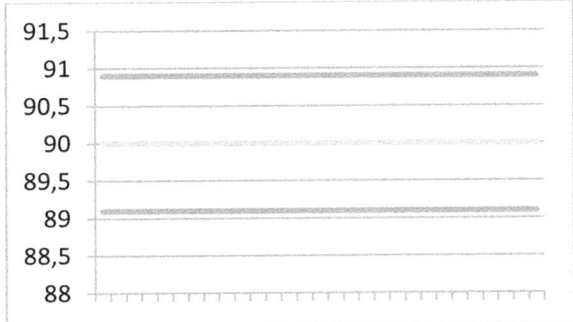

X-Axis: No n of produced doors
Y-axis: with in cm

Calibration compendium

As long as the mass of the door is between the two red lines, the door is within the promised / specified specifications.
This now has to be proven - that is to measure each door during production. A suitable measuring device is used for this.

This measuring device also has a basic tolerance. Cleverly, you choose a measuring device that clearly "better (m) i (s) st" than the permitted tolerance of the door.
As described in the "Measurement uncertainty" chapter, a measurement result is made up of a number of components - the tolerance of the measuring device is only one factor.

Suppose for the meter an expanded measurement uncertainty of ± 0.1 cm is shown - almost 10 times better than the permissible tolerance of the door - then a safe production should be guaranteed:

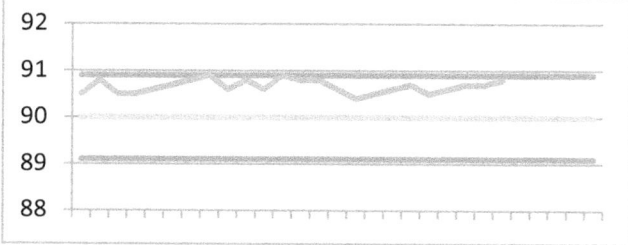

Calibration compendium

All doors produced are certainly "in tolerance" - some are irritating
maximum tolerance range - but are within the permissible limits,

If you now show error bars for the measurement uncertainty of the measuring device used, you can see:

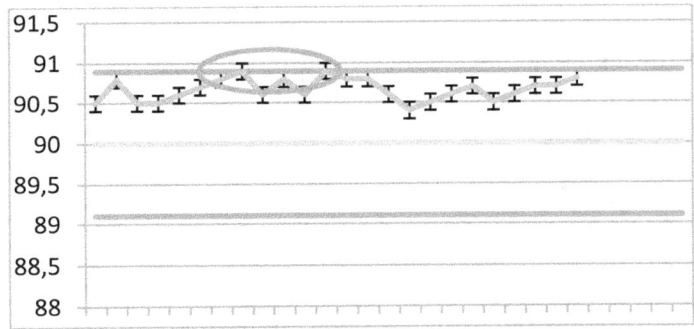

The doors located at the top of the tolerance, the statements "in tolerance" cannot be made with certainty!
Because the measurement result is uncertain, it could be possible that some of the doors produced are out of tolerance!
Here, the measuring method / measuring device or the specifications must be improved!

Calibration compendium

For the assessment of the suitability of a calibration, the following results:
- The product is not better or worse
- The evaluation of the product is not good enough for a qualitative assessment
- Either the scale is adjusted - or a calibration with smaller measurement uncertainty has to be requested!

Another view of tolerances and measurement uncertainty will be demonstrated using the already known targets:

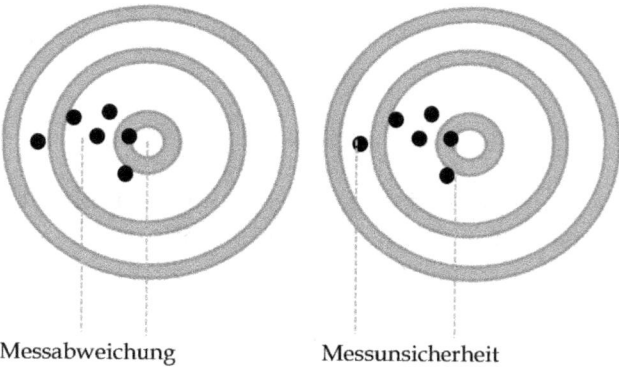

Messabweichung   Messunsicherheit

- The specification of a tolerance describes the (assured or desired) quality of a product.

Calibration compendium

- **The measurement uncertainty describes the quality of the measurement.**

When selecting a calibration service, it must therefore be assessed whether the quality of the calibration - expressed in terms of the measurement uncertainty, among other things (other features include the number of measuring points, direction, etc.) is sufficient for your own measuring device.

Calibration compendium

Calibration results

Calibration certificate

A report is included in the scope of each calibration - the calibration certificate.
The data that must be contained in a calibration certificate according to DIN ISO / IEC 17025: 2017.

According to DIN EN 17025, a calibration certificate must always contain:
- *a title (e.g. "Test Report", "Calibration Certificate" or "Report of Sampling") ;*
- *the name and address of the laboratory;*
- *the location of performance of the laboratory activities, including when performed at a customer facility or at sites away from the laboratory's permanent facilities, or in associated temporary or mobile facilities;*
- *unique identification that all its components are recognized as a portion of a complete report and a clear identification of the end;*
- *the name and contact information of the customer;*
- *identification of the method used;*
  a description, unambiguous identification, and, when necessary, the condition of the item;

Calibration compendium

- *the date of receipt of the test or calibration item[s], and the date of sampling, where this is critical to the validity and application of the results;*
- *the date(s) of performance of the laboratory activity;*
- *the date of issue of the report;*
- *reference to the sampling plan and samplingmethod used by the laboratory or other bodies where these are relevant to the validity or application of the results;*
- *a statement to the effect that the results relate only to the items tested, calibrated or sampled; the results with, where appropriate,*
- *the units of measurement;*
  additions to, deviations, or exclusions from the method;
- *identification of the person(s) authorizing the report;*
- *Clear identification when results are from external providers.*

NOTE *Including a statement specifying that the report shall not be reproduced except in full without approval of the laboratory can provide assurance that parts of a report are not taken out of context.*

Calibration compendium

Kalibrierlaboratorium für die Messgröße Drehmoment und Drehwinkel
Calibration laboratory for the measuring value torque and rotational angle

akkreditiert durch die / *accredited by the*
Deutsche Akkreditierungsstelle GmbH

als Kalibrierlaboratorium im / *as calibration laboratory in the*
Deutschen Kalibrierdienst

		0
Kalibrierschein	Kalibrierzeichen	D-K-123456-01-00
Calibration certificate	*Calibration mark*	2019-04

Gegenstand:	Drehmomentaufnehmer mit Messgerät	Dieser Kalibrierschein dokumentiert die Rückführung auf nationale Normale zur Darstellung der Einheiten in Übereinstimmung mit dem internationalen Einheitensystem (SI). Die DAkkS ist Unterzeichner der multilateralen Übereinkommen der European co-operation for Accreditation (EA) und der International Laboratory Accreditation Cooperation (ILAC) zur gegenseitigen Anerkennung der Kalibrierscheine. Für die Einhaltung einer angemessenen Frist zur Wiederholung der Kalibrierung ist der Benutzer verantwortlich.
Object	*torque transducer with measuring box*	
Aufnehmer / *Transducer* :		
Mod.Nr / *Mod.No* :		
Artikelnr. / *Art.No.* :		
Serien-Nr / *Serial number* :		
Hersteller / *Manufacturer*:	DearJohn USA	
Messgerät / *Measuring box*:		
Mod.Nr / *Mod.No.*:		
Artikelnr. / *Art.No.*:		
Serien-Nr / *Serial number*:		
Hersteller / *Manufacturer*:	DearJohn USA	
Auftraggeber:	Meier & Müller	*This calibration certificate documents the tractability to national standards, which realize the units of measurement according to the International System of Units (SI). The DAkkS is signatory to the multilateral agreements of the European co-operation for Accreditation (EA) and of the International Laboratory Accreditation Cooperation (ILAC) for the mutual recognition of calibration certificates. The user is obliged to have the object recalibrated at appropriate intervals.*
Customer		
	Gerstau 23	
	-42857 Remsheld	
Auftragsnummer	—	
Order No.		
Anzahl der Seiten des Kalibrierscheines:	5	
Number of pages of the certificate		
Datum der Kalibrierung:	2019-04-06	
Date of calibration		

Dieser Kalibrierschein darf nur vollständig und unverändert weiterverbreitet werden. Auszüge oder Änderungen bedürfen der Genehmigung sowohl der Deutschen Akkreditierungsstelle GmbH als auch des ausstellenden Kalibrierlaboratoriums. Kalibrierscheine ohne Unterschrift haben keine Gültigkeit.
This calibration certificate may not be reproduced other than in full except with the permission of both the Deutsche Akkreditierungsstelle GmbH and the issuing laboratory. Calibration certificates without signature are not valid.
This calibration certificate is based on the german language. In case of doubt only the german version is valid.

Datum	Stellv. Leiter des Kalibrierlaboratoriums	Bearbeiter
Date	*Vice head of the calibration laboratory*	*Person in charge*
2019-12-17		
	Armin Fuchs	Peter Jäger

Postanschrift/*Mail address* Telefon Durchwahl/ *Telephone extension*
DAkkS-Labor GmbH (+49) 0213) 123-4560
Kalibrierlaboratorium
Marina Str. 20
D-42937 Remscheid

Calibration compendium

In addition to these basic requirements, the standard specifies further requirements and devotes a separate chapter to these requirements with Section 7.8.3:

7.8.3.1 In addition to the requirements listed in 7.8.2, test reports shall, where necessary for the interpretation of the test results, include the following:
- information on specific test conditions, such as environmental conditions;
- where relevant, a statement of conformity with requirements or specifications (see 7.8.6);
- where applicable, the measurementuncertainty presented in the same unit as that of the measurand or in a term relative to the measurand (e.g. percent) when:
  - it is relevant to the validity or application of the test results;
  - a customer's instruction so requires, or
  - the measurement uncertainty affects
  - conformity to a specification limit;
- where appropriate, opinions and interpretations (see 7.8.7);
- additional information that may be required by specific methods, authorities, customers or groups of customers.

Calibration compendium

This still does not cover the requirements for a calibration certificate: section 7.8.4 also specifies special requirements for calibration certificates:

7.8.4.1 In addition to the requirements listed in 7.8.2, calibration certificates shall include the following:
- *the measurement uncertainty of the measurement result presented in the same unit as that of the measurand or in a term relative to the measurand [e.g. percent);*
  NOTE According to ISO/IEC Guide 99, a measurement result is generally expressed as a single measured quantity value including unit of measurement and a measurement uncertainty.
- *the conditions (e.g. environmental) under which the calibrations were made that have an influence on the measurement results;*
- *a statement identifying how the measurements are metrologically traceable*
- *the results before and after any adjustment or repair, if available;*
- *where relevant, a statement of conformity with requirements or specifications*
- *where appropriate, opinions and interpretations*

This section repeats and clarifies requirements that are essential:

The specification of the measurement uncertainty is mandatory - without the specification of the measurement uncertainty there is no calibration.

Calibration compendium

Intervals in a calibration certificate

A recurring requirement of the measuring device holder is to specify a calibration interval or to specify the time of the next calibration in a calibration certificate.

DIN EN ISO / IEC17025: 2018 states:

7.8.4.3 A calibration certificate or calibration label shall not contain any recommendation on the calibration interval, except where this has been agreed with the customer.

The measuring device holder (or the QM system of the measuring device holder) alone knows how the measuring device is used (single-shift operation or "around the clock", laboratory conditions or site use) and must incorporate these influences in the assignment of the calibration interval.
The specification of a calibration interval in a calibration certificate without the consent of the customer would mean an inadmissible interference of the calibration center in the QM system, the measuring device holder. This could even be an attempt at customer loyalty - this would not be permitted.

Calibration compendium

Therefore the interval specification must come from the measuring device holder; with this information - if it is preferably made known to the calibration laboratory in writing - consent is given.

DAkkS calibration marks never have an interval or the next calibration.

In addition to the DAkkS sticker, it is permissible to affix an (own) factory sticker with interval information or, better, the date of the calibration / the date of the next calibration to the measuring equipment.

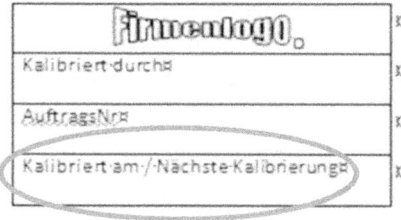

Calibration compendium

Statement of conformity

New in the DIN EN ISO / IEC17025: 2018 edition is the statement / specification about a declaration of conformity.

The inclusion of a declaration of conformity is based on the request of many measuring equipment holders and indicates whether a measuring device complies with specifications (e.g. the manufacturer's specifications) at the end of a calibration - or not.

In order to make such a decision, it has to be determined what the "rule" is: the norm speaks of a decision rule.

There is no binding requirement for a declaration of conformity. This can be agreed between the calibration body and the device holder. Therefore, there are different models that can be agreed:

Normative / zu vereinbarende Vorgaben zur Konformitätsaussage:
- Statement of conformity according to 14253-1
- Statement of conformity according to ILAC G9 8-2009
- Statement of conformity according to DAkkS-DKD-5
- Statement of conformity without taking into account the measurement uncertainty
- Statement of conformity according to individual customer requirements

Calibration compendium

Conformity statements / the decision rule according to customer requirements can be derived from this, e.g.

- without taking into account the measurement uncertainty
- "shared risk"
- individual requirements

An agreement must be made with the calibration center if a declaration of conformity is required in the calibration certificate.

The way of this agreement is currently practiced differently:

- The calibration center has a wording - if applicable also as a small footnote possible! – in the order confirmation
- The calibration center has a flyer / information letter that describes what the decision rule is, unless otherwise agreed
- A wording is contained in the general terms and conditions of the calibration center.

If a declaration of conformity in the calibration certificate is requested from the calibration center, a specification should also be made for the decision rule - otherwise the notification in one of the above or similar forms is considered an agreement!

Calibration compendium

The most common models of a decision rule are presented and explained below.

Decision rule - what does that mean?

The diagram shows possible cases of a measurement recording. The respective measured value lies on the dashed line, the double T represent the measurement uncertainty of the measurement.

At the first measuring point, the measured value including the assigned measurement uncertainty is clearly within the specification limit.

In the second case it can be seen that the measured value is clearly within the specification limit. However, due to the measurement uncertainty involved, the value could possibly also lie outside the specification limit.

In the third case it can be seen that the measured value is clearly outside the specification limit. However, due to the measurement uncertainty involved, the value could possibly also be within the specification limit.

Calibration compendium

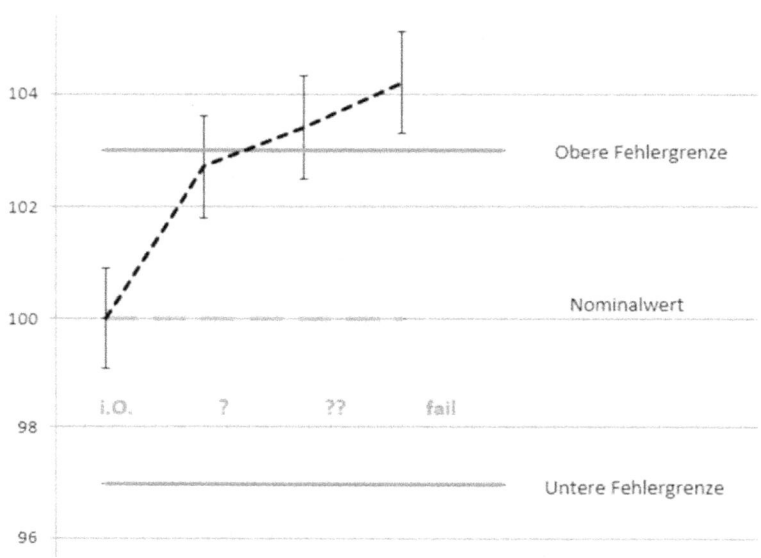

i.O.: The instrument complies with the specifications, taking into account the measurement uncertainty

?: The measurement is within the error limits. Taking into account the measurement uncertainty, no statement can be made about compliance with the specification. Compliance is more likely than failure.

??: The measurement is outside the error limits. Taking into account the measurement uncertainty, no statement can be made about compliance with the specification. Compliance is less likely than non-compliance.

Fail: The measurement result including the measurement uncertainty is greater than the upper error limit. The instrument does not meet the specifications.

Calibration compendium

As described above, the measurement uncertainty expresses the "quality" of the measurement or calibration.

Even if the measured value is "OK", ie within the specification limit, the actual result can be outside due to the uncertainty factor. The graphic shows that there are only two clear results: The first measuring point including measurement uncertainty is within the specifications - the last measuring point is clearly outside the specifications. This cannot be determined with certainty for the other two measuring points.

Calibration compendium

The ILAC G8 (ILAC: International Laboratory Accreditation Cooperation) implements these cases as "cases":

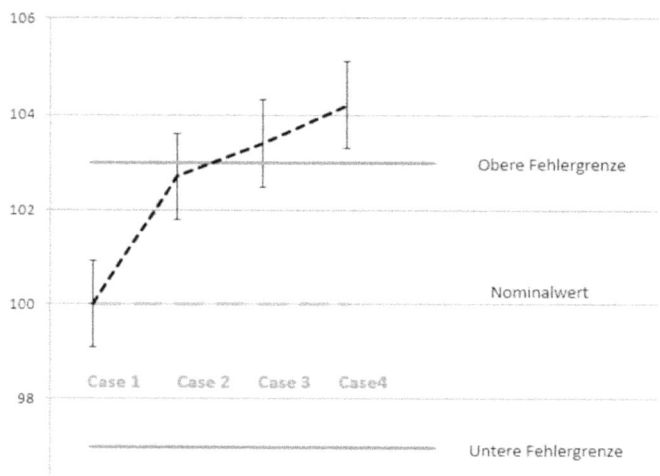

A clear statement on conformity or non-conformity can only be made in cases 1 and 4.

Calibration compendium

Must a device be "compliant" ?

The purpose of a calibration is to determine a measurement deviation. Ideally, a measuring device would always display exactly the value that it is measuring.

In practice, this is never the case.

But when is the device "OK" - when is it not?

All measuring devices have three characteristic basic features:
- Precision
- Repeatability
- Linearity

If a measuring device always measures "wrong" at the same point in its characteristic curve, this is not a problem: with a correction table - in modern devices by storing it in a memory block with automatic correction of the display - you can determine the correct value.

Calibration certificates indicate the values determined over the measuring range - the user can correct the measurement results.

This is too expensive or to much work for many users - there were demands on the standards committee to create an easy way of evaluating the calibration result. Users want to be able to quickly see whether the device is as precise as it has been when bought once - the declaration of conformity has been introduced.

Calibration compendium

However, there is still no reason not to work with the measured values determined during calibration - the device is not defective.

Calibration compendium

Need of test & measuring equipment

Measuring devices are all around - in every company and at home also. But why actually - for what purpose were they purchased? What should be considered when a measuring device needs to be replaced? If an offer for a calibration has to be obtained - what is the calibration center informed about the required calibration scope?

It would be easy to answer all of these questions if there were a type of specification for the existing measuring devices in which the specifications, measuring ranges and accuracy requirements are set. A data sheet is the specification sheet for the measuring device.

This data sheet can and will be in many points with a possibly available data sheet from the device manufacturer - nevertheless, the effort of own data sheets should be used to specify your own measuring tasks and the scope of calibration:

The device data in the manufacturer's documents also like to highlight company- and type-specific specifications or special features. Do you want e.g. Comparing the specifications of frequency counters from two manufacturers quickly reaches its limits if you want to compare beyond the typical core data such as frequency range or sensitivity.

Calibration compendium

Information is provided for manufacturer A, which manufacturer B searches in vain - and vice versa. Examples: Information about short-term - sometimes even "mid-term" - meaning medium term and long-term stability.

In addition, there are almost always special "options" or "features" with which the individual manufacturers try to differentiate themselves from others.
There are of course some sensible options - but some are to be seen as "nice to have" and not really necessary.

In the simplest case, a data sheet corresponds to the information provided by the manufacturer of the measuring device about its technical properties.

If you try to create a data sheet in which you simply "take over" this manufacturer's performance data, you quickly discover that the devices on the market almost always offer much more than is really needed. This can affect both the scope of services and the precision of a measuring device. This will be explained using the example of a multimeter.

Just over two decades ago, a multimeter was an analog display instrument with current, voltage and resistance measuring ranges.

Calibration compendium

The typical accuracy was +/- 3%. Such a multimeter was absolutely sufficient for many service tasks; calibration is not a problem.

With the advent of digital technology and a constant drop in prices, multimeters can today be purchased for very little money from numerous retailers - including cheap online providers.
The quality of these "instruments" should not be commented on here - good quality comes at a price.
But if you look at the performance data of these devices, you can see that you get significantly more performance than just a few years ago:
Usually a class better precision (smaller tolerances), many useful and just as much useless functions in addition: diode test, sometimes transistor test inputs, continuity tester, dB input and the like. One has to ask when buying whether all these functions are really needed!
Now the time has come for this multimeter to be calibrated: calibrating the full range of measuring device properties costs its price.

Means: either you have functions for expensive money calibrated that you don't need (because the original measuring tasks have remained about the same) or - which is unfortunately practiced too often, but blurring and watering down the qualitative status of your own business means you are looking for a DAkkS calibration center, gives the multimeter for

Calibration compendium

calibration and will soon receive it back with a calibration certificate.

Analyzing this calibration certificate, you will often find that the DAkkS laboratory has calibrated exactly what it is accredited for, e.g. Tension. But what about the other functions of the measuring device? Current, resistance range? The additional functions? So the multimeter is only somehow "half" and therefore incompletely calibrated. Only with the help of a calibration certificate can you see what exactly was calibrated.

If this device had a data sheet - one made by the device owner as described above - this data sheet would be a basis
- to find a matching calibration laboratory
- and thereby gain knowledge about the economic availability of this calibration resource.
- adjust the data sheet depending on the situation and thus gain a clear picture of your own actual needs
- and to be able to work economically in future

Calibration compendium

Such a data sheet is also the basis for an upcoming replacement. Because only the scope of services actually required may be the basis for a sensible and economical replacement - not what is "trendy" and which measuring device shines with great features.

A data sheet can be viewed as a specification and describes the properties that are really required for each measuring device.

This data sheet
- describes in detail what the measuring device "can" / must be able to do
- Is the basis for a buyer to find an equivalent or better replacement or successor device
- Is the basis for finding a suitable calibration center: when requesting an offer, add this data sheet and use it as the basis for the scope of calibration

Calibration compendium

When creating a data sheet, it should be recognized what the really required calibration scope / the really required parameter is:

Example: With a measuring bridge, a normal capacitor is inserted in a branch. This capacitor has a nominal value and an associated (manufacturing) tolerance and is classified as a single part that requires calibration.
The classification with the nominal value and the corresponding tolerance would be wrong.

In this case it would be correct to have a data sheet in which the nominal value is listed, but a statement "Deviation from the previous calibration max. +/- 2 nF.
It is not the absolute value that is important for this capacitor, but the variance / change in the current value since the last calibration.
The calibration variable was therefore not the nominal value, but the drift of this value.

This example is intended to show that values that are not "blindly" printed should be transferred to a data sheet, but that a data sheet should be a documentation of the performance parameters actually required.

Calibration compendium

Labeling of test & measurement equipment

ISO 9001: 2015 requests in section 7.1.5.2 ("Metrological traceability", extract):

„When measurement traceability is a requirement, or is considered by the organization to be an essential part of providing confidence in the validity of measurement results, measuring equipment shall be
   a) *calibrated or verified, or both, at specified intervals, or prior to use, against measurement standards traceable to international or national measurement standards; when no such standards exist, the basis used for calibration or verification shall be retained as documented information;*
   b) *dentified in order to determine their status;*

   c) *" … „*

The consequent implementation of this (considered very useful) requirement means:
Mark each measuring or testing device of your company with a calibration sticker, which must be labeled / filled in according to the entries of the measuring equipment monitoring.

Recommendation: Design a calibration mark for your company (also called calibration sticker or calibration label).

Calibration compendium

Such a uniform marking has a number of decisive advantages:

- Each employee can be instructed to always check the validity of the calibration stickers before starting their work.

- With a large number of measuring and testing devices, it is inevitable that measuring devices cannot be found temporarily or that they cannot be accessed. This can be done by internal relocations, by renting the equipment, by infrequent use, when it is handed over to field staff, etc. If such a measuring device reappears, the calibration status can be determined by briefly looking at the calibration sticker.

Calibration compendium

- If measuring and testing devices are assigned to external calibration facilities for calibration, you will usually receive there a calibration mark of the performing laboratory. This can be a DAkkS brand, but can also - e.g. for factory calibrations - be any brand of calibration device. The following applies to all of these calibration stickers: nothing is uniform. Different formats, colors, information mean that the content can only be poorly recorded and implemented by the user. If, in addition to these "third-party" stickers, a company-owned sticker is applied, this weak point is compensated for and the like. first list of lines can be emphatically requested.

A simple calibration sticker can also be obtained for little money. Even if there are no limits to the imagination, the number of fields should be limited to what is absolutely necessary - after all, the sticker should be able to be recorded "at a glance".

Calibration compendium

Example of a simple sticker:

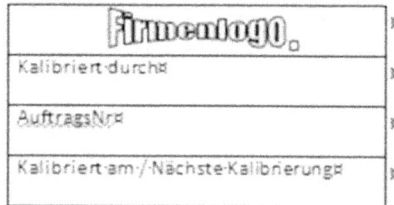

This only approx. 50 x 30 mm large sticker contains all the important information that
- establish references to the calibration center ("calibrated by", "order number") and the last calibration.
- state the current calibration status

Filling recommendation for such a calibration sticker:
Line 1:   Company name / logo, optional
Line 2:   Name of the calibration center, department if applicable
Line 3:   order number
Line 4:   "Next calibration" field:
            For calibrations and comparative tests: Specify the month with three letters, e.g. January, February, (exception: March), year
            For other measures: appropriate abbreviation (e.g. NCR, ICO)

Calibration compendium

With the specification of the calibration center and the order number
- If necessary, a new calibration can be requested (calibration laboratory can look up all the details, ranges, etc)
- the associated calibration certificate can be found *or reordered if lost*.

Each calibration center has an order or operational management system; an accredited calibration center is even obliged to keep calibration results (compare: ISO EN 17025: 2005, Section 5.4.7 "Control of data" and Section 5.9 "Ensuring the quality of test and calibration results").

The last job can be called up with the respective job number - all device data (type, serial number, etc.) are then available to the calibration center, and a new calibration can be prepared or offered. When "ordering" a calibration e.g. for a multimeter it may be the calibration for one of 10 multimeters in the company - the calibration center may calibrate several dozen multimeters of this type per day - the precise addressing of the multimeter is helpful here in order to obtain the desired scope and quality of calibration.

Calibration compendium

With repeated calibrations of the measuring device, the title pages of the calibration certificate show hardly any changes - holder data, type and serial number usually remain. unchanged. The creation date and the order number change. If you have a central or decentralized storage for calibration certificates, the last calibration certificate must be easily identifiable - the order number is helpful here.

If the calibration certificate is no longer available, a copy can be requested from the calibration center stating the order number.

Special attention should be paid to the 3rd line "Next calibration":
In principle, a calibration center may not specify the time of the next calibration:

ISO EN 17025: 2005, no. 7.8.4.3 "Calibration certificates":
A calibration certificate or calibration label shall not contain any recommendation on the calibration interval, except where this has been agreed with the customer.

This is set to prevent unwanted customer loyalty: it is basically the free decision of the device owner whether, when and how often he presents his device for calibration.

Calibration compendium

This free decision may be restricted by requirements such as those e.g. contain the Product Liability Act or ISO9001 - nevertheless, the device holder should remain free in his decision and not be bound to a calibration device.

On the other hand, with regard to these standards and laws, there is of course also the obligation to go below a calibration interval if the validity of the last calibration result has to be questioned due to failure, malfunction or break.

Having your own calibration sticker (which can be made available to a calibration center!) has several advantages:
- The agreement to specify a calibration interval is made by the provision,
- As explained above, every employee can be instructed to always check the (company) calibration stickers for validity before starting their work

Unfortunately there is no guideline for a standardized or standardized calibration sticker (yet). The information required above can only be read with difficulty or often not at all by many of the calibration stickers used.

Calibration compendium

Inconsistent versions, poor readability, information that appears to be arbitrarily necessary does not help and does not support modern measuring equipment management and does not meet the requirements of the standard for labeling.

With this diversity, it is impossible to expect that, as described above, the calibration status of the measuring device can be recorded "at a glance". An auditor will also find it difficult and immediately request calibration certificates.

Basically, you will not be able to dictate to any calibration body which sticker is required. This is remedied by the (company) own calibration sticker, which is stuck in place of or in addition to the calibration sticker of the calibration center. This is particularly recommended for DAkkS stickers; these should never be removed.

Calibration compendium

DAkkS calibration marks have a standardized form:

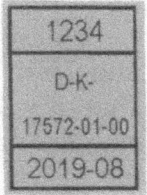

They are 18 x 28 mm in size and blue in color.
> Line 1: Count number
> Line 2: DAkkS registration number of
> the calibration center
> Line 3: year and month of calibration

With regard to further standardization, however, there are further efforts: here it is the military that has recognized through joint operations that standardization is essential to overcome language, structure and even writing barriers.

With the NATO paper STANAG No. 4704 "NATO Requirements for Calibration Support of Test and Measurement Equipment", (1st edition), a framework for a standardized sticker was defined.
As with many quality-related requirements from the military sector, it can be assumed that a corresponding norm or normative specification will also be implemented for the civil sector in a few years.

Calibration compendium

The calibration sticker shown as an example in this chapter already contains parts of the STANAG requirements.

Calibration stickers should be attached to the front or top of the device so that they are clearly visible. In the case of small measuring devices or measuring devices without space for a sticker, it can also be attached to the associated case / storage container (example: parallel gauge block: not every gauge block receives a sticker, the entire set bears a sticker on the usual wooden storage box).

Calibration compendium

Determination / adjustment of calibration intervals

There are numerous approaches for defining calibration intervals for measuring and testing devices:
- On the one hand rigid requirements to be applied,
- on the other hand, different approaches to the individual determination of a suitable interval. Here complicated mathematical relationships are often developed and stored for calculation.

An individual interval setting should only be made in exceptional cases. Basically, it is advisable to assign and provide uniform intervals for each type of measuring device:

In the case of a larger number of items and a variety of types of the existing measuring and testing devices, individual assignment of calibration intervals for each individual measuring and testing device should be avoided. This has the advantage that similar devices can be exchanged.

This is important with a holistic QM approach: if a technician fails his multimeter, it would be fatal if he borrowed the same type device from another department - but it was classified as not requiring calibration and, accordingly, was never calibrated.

Calibration compendium

Therefore it is strictly recommended:

Same type of device= same interval

As already described in a previous chapter, this interval is defined in a list of principles.
In the life cycle of the devices, however, it may be necessary to adjust the intervals: aging effects, knowledge of very stable or unstable devices, but also outliers ("Monday devices") may make such adjustments necessary.

First interval definition:
For a newly bought device, the manufacturer's specifications or recommendations should generally be adopted. If necessary, a different interval grid is selected, which allows practice-oriented calibration planning.
Depending on the device, the class must also be taken into account - a fine measuring manometer of class 0.1 should be recalibrated in a much shorter time interval than a manometer of class 2.5.

Calibration compendium

Interval adjustment

The calibration interval is gladly increased for cost reasons. You should only do this
- if the measuring device allows such an interval change in terms of design.
- An evaluation of the calibration results allows this.

Long interval due to the device design:
There are a few measuring and testing devices that have practically no mechanically movable components and are therefore not subject to mechanical wear.

Examples:
- Triple point cell water
- Fix point cells tin, zinc, silver, gold
- Flow meters based on the Coriolis principle

Such devices may have a long calibration interval of e.g. 4 - 5 years. However, it is wrong to believe that there is no change in such measuring devices: triple point cells water e.g. consist of a glass body that is filled with technically pure water and inside which a certain pressure prevails. There is no outside influence on the inside of the cell.

Calibration compendium

Nevertheless, such a cell also drifts: over time, the glass releases ions into the water and "contaminates" it - the plateau temperature shifts minimally.

Evaluation of calibration results:

If you want to make an interval adjustment based on the technical device history, you have to deal with the results of calibrations. This means that the calibration results - usually documented on calibration certificates - must be evaluated.

A simple but very effective method is the assignment of processing codes after calibration or after every repair / repair:

>Example 1:
>C: Calibration
>A: Adjustment
>R: Repair

Or:

>Example 2:
>C: compliant
>N: non compliant
>A: Adjustment
>R: Repair

Calibration compendium

The form and degree of differentiation of such a coding are almost unlimited and can be set for your own company - you should keep the codes manageable and think of options for later evaluation.

One way of evaluating interval adjustment and individual evaluation is presented and explained below:

Why the above Code Assignment? In the diversity of all measuring devices, real comparability is impossible. A standardized evaluation is also impossible - how should the calibration results of a spectrum analyzer with possibly ten thousand automatically recorded measurement data be compared with the calibration result of a spring balance?

Even if only similar devices have been turned in for calibration for calibration - for example a number of multimeters - the evaluation of the calibration results by comparing the measurements is tedious and time-consuming.

However, if you ask e.g. Basically a statement of conformity from the calibration device, one knows with little effort whether or not the measuring device complies with the manufacturer's specifications or the specifications in the data sheet that was used as the basis for the calibration.

Calibration compendium

The assignment of a code letter is therefore easy - this code letter is entered in a measuring device database.

This measuring device database with all order data and processing codes allows the reliability data of the devices to be evaluated
- holistic
- type-related or
- serial number-related.

The CV database can be used to assess whether
- all devices of one type are unstable -> shortening the intervals
- individual devices are unstable -> discarding
- Devices are stable -> extending the interval

Individual decisions are possible, measured values can be used as a support

By evaluating simply the code letters, you can quickly identify the metrological trend of the respective measuring device. Of course, a minimum number of calibrations is required - fewer than three separate calibration results should not be available.

Calibration compendium

Prolongation of calibration interval

If the two cases described above do not apply, there is no technically justifiable decision to extend the interval.

A purely economic decision is strongly discouraged. Also the popular argument "measuring device is used very little / rarely does not stand up to a professional test:

Measuring devices that contain mechanical components also experience changes when not in use, which have a significant impact on their metrological properties.

Examples:

Measuring devices with moving components such as torque wrenches:
Internally used greases can become gummy, bearings become stiff, springs change their constant depending on the tension.

Measuring devices without moving components such as mass pieces:
Can, e.g. change their buoyancy due to low magnetic influences; the surface can oxidize when stored in non-air-conditioned rooms.

Electronic measuring devices in general, such as multimeters, oscilloscopes, etc.: Components (capacitors e.g.) can dry out.

Calibration compendium

If a calibration interval is extended extremely - e.g. for 4 years (this interval is actually encountered in practice), a warning must be given before using such measuring and testing devices - they must be regarded as not calibrated.

It also makes no sense to extend calibration intervals (significantly) because the measuring device is not required.
Better: Block the measuring device with a clear identification on the device (calibration sticker), deny access ("include") and make a corresponding entry in the calibration overview.

Calibration compendium

Intervals in a calibration certificate

A calibration laboratory is generally not allowed to specify an interval. This is to avoid an (economic) dependency of the customer on the calibration laboratory.

However, an agreement with the client / customer is permitted; If the customer requests an interval or an appointment for the next due calibration, this may also be noted in the calibration certificate.

However, after extreme intervals such as the above-mentioned 4 years are actually encountered, various working groups of the DKD / DAkkS specialist committees have been set up (working groups for the development of DIN or VDE / VDI guidelines) decided to limit the validity period of calibration certificates: the calibration certificates have a validity period, not the "calibration" or the calibration interval!

Beispiel:
DIN 51309: 2005-12 writes in paragraph 6.3.2 recalibration:
"The calibration certificate is valid for a maximum of 26 months."
Here you can see the clear recommendation of a two-year interval. In order to be able to compensate for unpredictability (unavoidable measurement requirements, audit, appointment of the calibration center, etc.), 2 more months were added.

Calibration compendium

Indication of the time of recalibration

A question that is often discussed is: When exactly does the calibration expire? (See also the above statements on "interval specifications in the calibration certificate").
There is no official or standard-based requirement to answer this question.
Has a measuring device experienced a calibration interval of 12 months and the last calibration on April 16, 2019 - when does it have to be calibrated again?
There are operational management systems that are also used for measuring equipment monitoring that keep a complete date.

Accordingly, the measuring device should no longer be used from April 17, 2020 unless it was recalibrated beforehand.

It would be better - because it is more practical - to make a stipulation that a measuring device must always be presented for calibration on a monthly basis. In our example this would be in April 2020.

This means that a maximum of almost a full month can be "won". At first glance, this only appears as additional time of use. As a rule, the re-presentation for calibration cannot be coordinated to the day - here is admitted many imponderables that can be in your own company, but can also be with an external calibration service provider.

Calibration compendium

Choosing the "target month" as the expiry date therefore only gives the necessary scope for everyday calibration registration.

If you decide to do this,
- It must be recorded in the quality management manual
- The re-presentation date must be kept in the measuring equipment monitoring accordingly
- The own calibration sticker (if used) shall only be filled with the specification of the month and year.

Calibration compendium

Start of a calibration interval

The expiry of a calibration interval begins to count from the date of the calibration. It is a common misconception that time only runs from the first use.

It is certainly annoying that part of the period of use has to be allocated to logistical processes such as shipping. But it must be understood that it is not the use of a measuring device that influences many factors. This also includes influences that can have negative effects on the measuring device during storage or "non-use".

Examples are already listed in the "No interval extension" section.

Interruption of usage

Also an interruption in the use of a measuring device and e.g. its storage for a period <u>does not prolong</u> that calibration interval. The interval begins when a calibration is completed.

Calibration compendium

General information about calibrations and calibration intervals

About 8% of all measuring and testing devices must be set and adjusted during a calibration.

However, a clear distinction must be drawn:
If a device is presented to the manufacturer or a calibration device (often identical) as "defective", it does not count for this 8%.
The device holder knew that there was a failure (failure, malfunction or break). Devices are taken into account that should "only" be calibrated and which the user did not assume to be measuring incorrectly or inaccurately.

Although this figure has a certain uncertainty, it is based on an evaluation of a large device pool (> 200,000 devices, over 10,000 different device types of all types: from simple multimeters to spectrum analyzers, from coarse scales to low-pressure standards); coincides with the experiences of other device holders with a large amount of measuring and testing devices and can serve as a "thumbs-up value" and thus as a guideline for your own target of a quality score.

Calibration compendium

Calibration planning / scheduling / delivery

Factory or traceable calibration

At the latest when planning / registering for calibration, you have to ask yourself whether a factory calibration is sufficient or whether a DAkkS calibration is required (cf. also "List of principles").

A common - but inconsiderate - approach is often the question of cost. The answer can be given as a blanket: a factory calibration is always significantly cheaper. This often makes the decision for this calibration.
Such an approach is too short and often becomes a problem during the next audit.

There is no official definition for the term factory calibration. At first you don't really know what is on offer. A factory calibration is usually carried out based on standards and regulations or based on customer requirements.

This may be sufficient for many applications, but the decision for a factory calibration should be well thought out and then based on a decision matrix.

Calibration compendium

The German DAkkS paper "DAkkS-DKD 4" defines factory calibration as follows:

"In-house calibration (factory calibration)

6.4.1 An in-house calibration system ensures that all measuring and testing equipment used in a company is regularly calibrated with the company's own reference standards. For the reference standards of the company, the traceability of the measurements must be ensured by calibration in an accredited calibration laboratory or a metrological state institute. The in-house calibration can be verified by an in-house calibration certificate, a calibration mark or another suitable method. The calibration documents must be kept for a prescribed period.

6.4.2 The type and scope of the metrological control during an in-house calibration are left to the company concerned. They must be adapted to the special applications so that the results achieved with the measuring and testing equipment are sufficiently accurate and reliable. Accreditation of the organizations that carry out internal calibrations is not necessary to meet the requirements of the EN ISO 9000 series of standards applied to internal requirements. However, if an external body uses an in-house calibration certificate as proof of return, it should be required that the issuing organization can demonstrate its competence."

Calibration compendium

An analysis of this text shows for the implementation:
- A factory calibration certificate is regarded as an internal calibration certificate.
- If the factory calibration certificate of another, external calibration body is to be recognized / used, this body should be accredited.

When choosing a suitable calibration laboratory for a factory calibration, a calibration body that is accredited in at least one parameter according to EN 17025 is the first selection criterion:

A calibration laboratory that has this type of accreditation must have a quality management system and, manadtory, observe the entire range of the requirements of this standard.

Although it is possible, it is extremely unlikely that a laboratory that requires accreditation will have a parallel system, which leaves all organizational and infrastructural expenses behind and carries out a "backyard calibration" without any quality-related requirements.

The second selection criterion should be the determination of what is measured with the measuring and testing device. In this case, it is not the measurement parameters that are meant, but the areas of application of the instruments.

Calibration compendium

The following questions should be answered:

- Is the measuring device a device for daily use (occasional measurements: e.g. a slide gauge to determine the drill thickness, multimeter to determine the voltage supply of a socket, etc.)
- Or are regular, critical or quality-relevant measurements carried out (e.g. re-measuring of production dimensions)?
- Or is the measuring device even calibrating (checking and calibrating other company measuring devices, e.g. multimeter or torque wrench)?

Calibration compendium

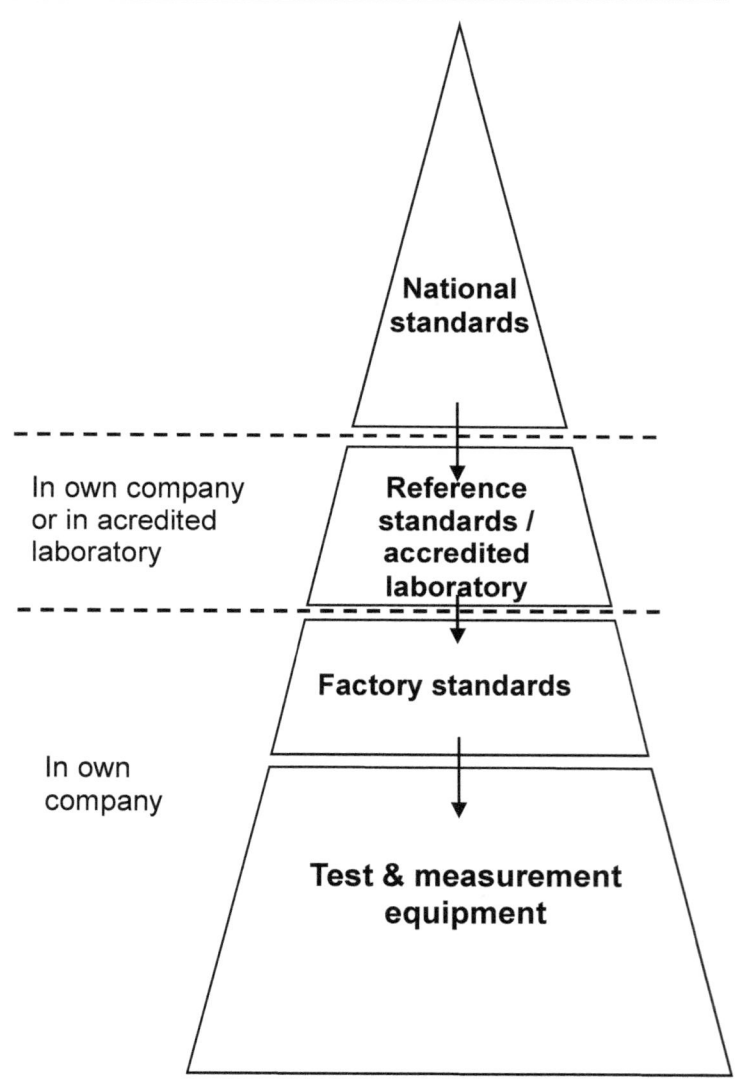

Calibration compendium

To answer these questions, the measuring and testing devices should be grouped according to their area of application.

This grouping is easiest to use the pyramid already presented to visualize your own calibration hierarchy.

Arrange (the relevant) measuring and testing devices in the corresponding levels of the pyramid - usually only the two lower levels come into question, level three can occur with individual companies.

If you examine the measuring and testing devices of your own company, you may find that there may even be own factory standards.

Examples:
- A calibration device (example: SCHATZ caliTest) is available to be able to calibrate the torque wrench.
- To be able to calibrate the multimeter, a calibrator (example: FLUKE 5101) is available

The DAkkS-DKD-4 provides precise information here: *"Usage standards or factory standards must be calibrated by an accredited calibration laboratory and thus be traceable to national standards."*

This provides clear decision support.

Calibration compendium

Summary: The difference between DAkkS calibrations and factory calibrations can be that applicable standards or regulations for calibration (e.g. VDI / VDE 2646 for torque sensors) are only used in a simple, "slimmed down" form. This enables simple, inexpensive and technically correct calibrations to be carried out.

In the case of DAkkS calibrations, the entire procedure for a calibration, including the form and content of the calibration certificates, is prescribed by the DAkkS / DKD, which means that these calibrations are considerably more complex.

There are no specifications for all parameters for the creation of factory calibration certificates and is then within the discretion of the calibration laboratory.

In the case of factory calibration certificates, it is therefore imperative to ensure that the measurement uncertainty and the proportions that lead to the measurement uncertainty are shown.

Calibrations without specifying the measurement uncertainty are worthless.

A calibration hierarchy decides whether factory or traceable calibrations are necessary, which should be created for operation.

Calibration compendium

Measuring chain or individual devices

The international dictionary of metrology defines a measuring chain as follows:

"Measuring chain - sequence of elements of a measuring system that forms a single path of the signal from a sensor to an output element."

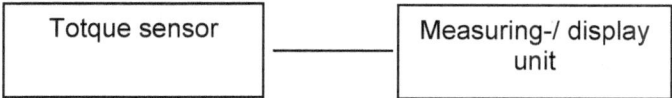

If you are in the situation of having to decide whether to present individual components or an entire measuring chain for calibration, you should consider:

Calibrated as a measuring chain, the example measuring chain shown - which ideally was calibrated together - has an overall measurement uncertainty x.
However, if the components were individually calibrated, these calibrations would result in the individual measurement uncertainties y and z.

The total measurement uncertainty must now be calculated by yourself: As described in the "Measurement uncertainty" chapter, in the simplest case a geometric addition multiplied by the extension factor k can be:

$$u = \sqrt{y^2 + z^2} * k$$

Calibration compendium

The resulting expanded measurement uncertainty is usually greater than the measurement uncertainty x that is achieved with a measurement chain calibration.

If this measurement uncertainty "fits" in your own measurement uncertainty budget, a calibration of the individual components can be sensible or more economical.
However, it is generally not advisable to separate measuring chains.

Calibration compendium

Failure of a measuring system / repairs

What to do if a measuring and testing device fails within the calibration interval?

A repair is not enough here - a calibration must then be carried out.

Even if on a larger test stand e.g. "only" the power supply fails and no sensors (e.g. sensors or simulators) are affected, a holistic calibration is required:

Such a test bench is a complex system - even individual settings on the power supply can have an impact on the measurement accuracy and thus measurement uncertainty of the system.

Only a calibration after the repair ensures that a test bench is true to size and ready for use.

A partial calibration (e.g. after the ohmic range of a multimeter has failed) is also not advisable - no one can rule out the effects of the defective measuring range on the rest of the measuring device without a check (which is equivalent to a calibration).

Calibration compendium

Calibration: Laboratory, „on-site" or „in-situ"

As far as technically possible, calibrations are carried out flexibly and in accordance with customer requirements and taking the formal requirements into account.

- At the end of the manufacturing chain, the first calibration takes place in the factory ("end of line calibration").

- Recalibrations are carried out in a calibration laboratory either as traceable calibration or as service calibrations (factory calibrations).

- Service calibrations are often offered on site at the customer's site and traceable calibrations are also available in many parameters. Before commissioning a calibration service provider, it is necessary to inquire or check whether this service provider is also accredited for on-site calibrations.

- In some cases, only an "in-situ" calibration is possible: wherever the measurement setup must not be dismantled due to specifications (e.g. in medical technology). As a rule, special certification or approval is required.

Calibration compendium

Selecting a calibration laboratory

After making the decisions regarding measuring chain calibration or calibration of individual components, it is now a matter of selecting a suitable calibration laboratory and registering the measuring and testing device for calibration.

A simple submission for calibration without instructions or agreements can correspond to the presentation of a motor vehicle in a workshop with the request for inspection: what exactly should be done? What can or may be included in the scope of services? The idea of calibration should not be a license for measures that the calibration laboratory deems necessary, but must be tailored to the requirement profile of the contracting company and the respective measuring device.

We recommend submitting the data sheet described above as the basis for a calibration. This data sheet can now be the basis for the requirements of the calibration body with regard to its scope of services - it can (e.g. based on the representation of the scope of accreditation Internet on the DAkkS homepage: http://www.dakks.de/content/akkreditierte-stellen-dakks) are checked whether the calibration center fulfills the required scope of services of the data sheet.

Calibration compendium

A multimeter e.g. is turned in for for DAkkS calibration, so the typical expectation is to get a fully calibrated measuring device back. In the case of a multimeter, this would be at least voltage, current, resistance - in modern devices also temperature, diode test, etc..

Such an "all-round calibration" is not economically sensible in many cases: if the multimeter is e.g. used only for the measurement of voltages, all other parameters are calibrated without benefit but at rather high costs. The range of services that is actually required on the data sheet helps to save these unnecessary costs.

Such a definition also helps with the chosen calibration laboratory: a calibration laboratory that is accredited for voltage is not necessarily also accredited for current or resistance. It can happen that a DAkkS calibration certificate is issued for the voltage range, and only a factory calibration certificate for the other measuring ranges.

The decision whether this is sufficient should be made before submitting for calibration and order placement.

Calibration compendium

Class of calibration

Depending on the type of measuring device, it may happen that different calibration qualities / classes are offered.
This is often the case with measuring devices that can be subdivided into classes. Here it must be analyzed which class applies to your own measuring device. The class imposes the measurement uncertainty requirement. Many calibration devices provide staggered prices for calibration by class - here you can save if you choose the right one.

For on-site calibrations - i.e. calibrations in which the calibrator comes into operation - it must also be researched beforehand which measurement uncertainty can be achieved on site. This can be done by inquiring to the calibration office - but it is better to inquire about the scope of accreditation, which can be queried on the DAkkS homepage at
http://www.dakks.de/content/akkreditierte-stellen-dakks).
With accredited calibration laboratories, the best measurement uncertainty that can be achieved on site is not always the same as the measurement uncertainty that can be achieved in the laboratory.

Calibration compendium

After a calibration / receiving your instrument

If you get the measuring device back after calibration, you should urgently check it. In a calibration laboratory, metrological, but also e.g. formal errors arise - here too "only" people are at work - it is advisable to draw up a checklist:

Calibration certificate:
- Has a calibration certificate been delivered?
- Is the information in the calibration certificate correct?
- Device data such as serial number
- Order number
- Customer / client: If a service provider is commissioned with the monitoring of measuring equipment and also commissions the calibration center, the name of the service provider must not be included in the calibration certificate: the name of the device holder must be listed.
- Does the scope of calibration documented in the calibration certificate match the assignment?
- Is the calibration certificate plausible in terms of measurement technology? Are the measuring range and / or measuring directions correct?
- Are the data for recalibration specified and still valid when listing the standards used?

Calibration compendium

Checks on the device:
- Has a calibration mark been applied?
- Are the entries on the brand correct?
- Are the returned accessories (power cord, rack mount kit, manual, adapter, etc.) complete?
- Has an electrical safety test been carried out and documented (depending on the device type and regional laws and regulations)?

The most important step is the evaluation of the calibration certificate:
It has to be checked
- whether the measuring and testing device is within the tolerance
- if necessary, a comparison has been made.

If it turns out that the measuring device was / is out of tolerance in one or more parameters, measures must be initiated immediately. This can be the recall of products that have been measured or monitored with the affected measuring device or a reworking of workpieces.

Good and responsible calibration facilities send a message to the device holder if irregularities are found.

Calibration compendium

In case that a measuring and testing device is found to be "out of tolerance", a chapter on the measures to be applied should be included in the quality management manual.

A distinction should be made for the sake of completeness: The above points only apply if a device is given for calibration in good faith that it is OK.

If a device fails due to failure, malfunction or breakage, you can make appropriate decisions immediately - a calibration certificate after the repair has been carried out should not be waited for.

Calibration compendium

Machine capability

The term "machine capability" comes from production technology and has nothing! - to do with calibration!

In the field of industrial screw connections, however, inquiries for implementation are also made to calibration service providers for torque or angle of rotation.

The term "machine capability" refers to a process that can be examined and proven whether a machine can be produced with sufficient security against errors. The automotive industry often requires proof of an MFU and the achievement or exceeding of specified parameters. These characteristic values are the two indices / key figures Cp and CpK.

For example, in order to investigate the influence of a screwdriver on the manufacturing process, 50 measured values are carried out without interruption under optimal conditions using a reproducible measurement setup and recorded with a suitable reference system.

Then the machine capability index (Cmk) is determined depending on the required tolerance values (limit values) and the standard deviation determined.

Calibration compendium

The 5-M influences count as factors that have a significant influence on the test result: people, material, measuring method, machine temperature and manufacturing method.

These may not change or may change only slightly.

- Man: The same person must operate the machine during the exam.
- Material: The same material must be used.
- Measurement method: The same measuring device is used for the entire duration of the examination.
- Machine temperature (environmental influences) : The temperature of the machine should not fluctuate and has reached operating temperature.
- Manufacturing method: The same manufacturing method (procedure) is used.

After completing the series of measurements, the statistical distribution applicable to the sample is determined. A normal distribution can generally be assumed, but in practice more complex methods are used.

Calibration compendium

After determining the position and scatter of the measured size, the machine capability can be determined as a numerical value.

A machine capability of 1.33 (corresponds to 4 σ standard deviation with normal distribution) or 1.67 (corresponds to 5 σ) is often used. The smaller the value, the poorer the machine capability.

Calibration compendium

Measurement system analysis MSA

The term measuring equipment capability analysis, also called measuring system analysis or measuring equipment capability analysis, MSA is a derived term from the Six Sigma (6σ), a management system for process improvement, statistical quality goal and a method of quality management.

These terms stand for the analysis of the capability of measuring equipment and complete measuring systems.

These terms also have nothing directly to do with calibration. However, they always build on (traceable) calibrated measuring devices - this is essential in any case.

Rather, a measuring device capability must be considered before and after the acquisition of a measuring device: If measuring devices are integrated directly into the production process for continuous monitoring, it should be checked when designing this production system whether the intended measuring devices are suitable for the application.

Calibration compendium

The basic properties of each measuring device described at the beginning of this book

- Accuracy, trueness, bias)
- Repeatability
- Reproducibility
- Stability
- Linearity

Type-1 study

This procedure examines the accuracy and repeatability of a measuring system. For the examination, 25 to 50 times are measured against a standard. The indices Cg and Cgk are then calculated based on the standard deviation of the measured values and the systematic measurement deviation. The calculation steps correspond to those of a process capability analysis. This is the typical procedure with which the suitability should be determined BEFORE (first) use or purchase.

Calibration compendium

Type-2 study, Gauge R&R study

This procedure examines the repeatability and comparative precision of a measuring device and is usually used as a repeat test (as a delimitation to the test to determine suitability "study 1) e.g. also used before or during audits ..

The name comes from the English "repeatability and reproducibility", hence R&R or also Gauge R&R or Gage R&R (English ga [u] ge = measuring device) and is only used when the measuring device has been classified as capable according to method 1 ,

Ten test objects, which should cover the entire range of the measured feature, are measured two or three times by three different operators (or at three different locations or with three different devices of the same type). The results are to be kept closed for each operator and the other operator is to be seen. The test objects should be measured in a random order in order to avoid the transfer of results ("recall memory") from the previous measurement.

When the series of measurements is complete, an overall mean value and an average span value (based on the differences between the largest and the smallest measured value that the operator has determined for each part) are calculated for each operator.

Calibration compendium

The difference between the largest and the smallest operator mean value allows a statement about the comparative precision; the total mean value of the average span values calculated for the individual operators is used to make a statement about the repetition precision. On the basis of repetition and comparison precision, the total scatter of the measuring device is then calculated and related to the scatter of characteristics or tolerance.

Type-3 study, R&R study

This method 3 is a special case of method 2, which assumes that the operator cannot influence the measuring device or the influence is negligible and is used above all in automated measuring systems.

Even if the basic systematic corresponds to the tests described above, there is no "standard measuring system analysis". There are implementing regulations for almost every area, also designed as VDE / VDI sheets.

The automotive industry has developed and published a guideline "Proof of capability of measuring systems". This generally applies as a guideline for the determination of measuring equipment capability.

Calibration compendium

Bibliography

Bibliography

Source literature / Sources used and cited:

Burghart Brinkmann
„Internationales Wörterbuch der Metrologie - Grundlegende und allgemeine Begriffe und zugeordnete Benennungen (VIM)"
Beuth Verlag

PTB, Physikalisch-Technische Bundesanstalt
Nationales Metrologieinstitut
PTB-Infoblatt, 2017
„Das neue Internationale Einheitensystem (SI)"

PTB, Physikalisch-Technische Bundesanstalt
Nationales Metrologieinstitut
„Die gesetzlichen Einheiten in Deutschland"

DIN EN ISO 9001:2015
„Qualitätsmanagementsysteme Anforderungen"
Beuth Verlag

DIN EN ISO 10012:2004-03
"Messmanagementsysteme - Anforderungen an Messprozesse und Messmittel"

DIN EN ISO / IEC 17025:2018
"Allgemeine Anforderungen an die Kompetenz von Prüf- und Kalibrierlaboratorien"
Beuth Verlag

Bibliography

DIN EN ISO 19011:2011-12
" *Leitfaden zur Auditierung von Managementsystemen*"
Beuth Verlag

DIN 51309:2005-12
„Werkstoffprüfmaschinen– Kalibrierung von Drehmomentmessgeräten für statische Drehmomente"
Beuth Verlag

DIN ISO 55350-11:2008-05
"*Begriffe zum Qualitätsmanagement - Teil 11: Ergänzung zu DIN EN ISO 9000:2005*"
Beuth Verlag

Deutscher Kalibrierdienst, Richtlinien / Leitfäden
"*Rückführung von Mess- und Prüfmitteln auf nationale Normale*"
www.dkd.eu

Bernd Pesch
"*Messunsicherheit: Basiswissen für Einsteiger und Anwender*"
Books on Demand

Peter Jäger
„Messmittelmanagement und Kalibrierung"
Books on Demand, ISBN 9783750434189

www.ingramcontent.com/pod-product-compliance
Lightning Source LLC
Chambersburg PA
CBHW050218230526
45470CB00001B/435